PENGUIN BOOKS

CHILDREN OF THE FLAMES

Lucette Matalon Lagnado was educated at Vassar College and Johns Hopkins University School for Advanced International Studies. The story of the Auschwitz twins is one she has covered since 1984. A former associate of syndicated columnist Jack Anderson, Lagnado also was an investigative reporter for the *New York Post* and *The Village Voice*. She is now Executive Editor of *The Forward*, a national weekly Jewish newspaper. She lives in New York City.

Sheila Cohn Dekel is a writer, lyricist, and educator. She is the widow of Alex Dekel, whose story is recounted in this book. She collaborated with him on many projects, including his long pursuit of Josef Mengele. She lives in New York City and is currently finishing her first novel.

CHILDREN OF THE FLAMES

DR. JOSEF MENGELE AND THE UNTOLD STORY OF THE TWINS OF AUSCHWITZ

Lucette Matalon Lagnado
&
Sheila Cohn Dekel

PENGUIN BOOKS

PENGUIN BOOKS
Published by the Penguin Group
Penguin Group (USA) Inc., 375 Hudson Street, New York, New York 10014, U.S.A.
Penguin Group (Canada), 90 Eglinton Avenue East, Suite 700, Toronto,
Ontario, Canada M4P 2Y3 (a division of Pearson Penguin Canada Inc.)
Penguin Books Ltd, 80 Strand, London WC2R 0RL, England
Penguin Ireland, 25 St Stephen's Green, Dublin 2, Ireland (a division of Penguin Books Ltd)
Penguin Group (Australia), 250 Camberwell Road, Camberwell,
Victoria 3124, Australia (a division of Pearson Australia Group Pty Ltd)
Penguin Books India Pvt Ltd, 11 Community Centre, Panchsheel Park, New Delhi – 110 017, India
Penguin Group (NZ), cnr Airborne and Rosedale Roads,
Albany, Auckland 1310, New Zealand (a division of Pearson New Zealand Ltd)
Penguin Books (South Africa) (Pty) Ltd, 24 Sturdee Avenue,
Rosebank, Johannesburg 2196, South Africa

Penguin Books Ltd, Registered Offices: 80 Strand, London WC2R 0RL, England

First published in the United States of America by
William Morrow and Company, Inc., 1991
Reprinted by arrangement with William Morrow and Company, Inc.
Published in Penguin books 1992

36 37 38 39 40

THE LIBRARY OF CONGRESS HAS CATALOGUED THE HARDCOVER AS FOLLOWS:
Lagnado, Lucette Matalon.
Children of the flames : Dr. Josef Mengele and the untold story of
the twins of Auschwitz / Lucette Matalon Lagnado and Sheila Cohn Dekel.
p. cm.
Includes bibliographical references (p.)
ISBN 0-688-09695-6 (hc.)
ISBN 0 14 01.6931 8 (pbk.)
1. Mengele, Josef, 1911– . 2. War criminals—Germany—
Biography. 3. Physicians—Germany—Biography. 4. World War,
1939–1945—Poland—Atrocities. 5. Human experimentation in
medicine 6. Twins—Biography 7. Holocaust, Jewish (1939–1945)—
Poland—Personal narratives 8. Auschwitz (Poland : Concentration
camp) I. Dekel, Sheila Cohn. II. Title.
DD247.M46M38 1990
90.54'05'092—dc20 90–13387

Printed in the United States of America
Designed by Ruth Koibert

To my mother, Edith Matalon Lagnado,
and to my doctor, Burton J. Lee III, M.D.,
who both saved me, each
in their own way
—L.M.L.

To the six million—and one . . .
—S.C.D.

Preface

CANDLES IN THE NIGHT

In the winter of 1984, I was asked by *Parade* magazine to seek out the long-lost child survivors of Dr. Josef Mengele's experiments at Auschwitz during World War II. Of the estimated three thousand twins—most of them young children—who had passed through Mengele's laboratories between 1943 and 1944, only about a hundred were known to have survived. Many fewer were thought to be still alive when I undertook my search for them in America and Israel.

I was aided in this effort by Eva and Miriam Mozes, twin sisters who had undergone painful experiments at Mengele's hands. Eva, an American housewife living in Terre Haute, Indiana, had long believed the twins' ordeal, which had received no more than a passing reference in the history of the Holocaust, should be told. Determined to reunite all the surviving twins, Eva and her sister had just founded an international society of Auschwitz twins—CANDLES—an acronym for Children of Auschwitz Nazi Deadly Laboratory Experiments Survivors. Through this group, the twins hoped to piece together what had been done to them in Mengele's laboratory. For unlike other death-camp survivors, these child victims had kept totally silent about their past. Ridden with guilt and with shame, most had never breathed a word about their ordeal as guinea pigs of the notorious Dr. Mengele, not even to loved ones. Theirs was the untold story of the Holocaust.

A few well-placed ads in Israeli newspapers yielded gratifying results. Within days, dozens of twin survivors (most of whom had ended up in Israel after the war) came forward to join CANDLES. In tearful reunions, they reforged the bond that had helped them survive both the death camp and Mengele's demonic experiments. All were united in their fierce desire to see the war criminal brought to justice.

I arrived in Israel in March to begin my interviews with Mengele's twins. Although, thanks to CANDLES' burgeoning membership rolls, tracking down their names and addresses was relatively simple, arranging actual meetings with them was considerably more difficult. Only at the gentle prodding of Miriam and Eva did a number of twins agree to see me. Each session began with the whispered confession, "I have never spoken of this to anyone."

But for many of the twins, simply telling what happened to them proved to be a release. Middle-aged, they spoke with the candor, intensity, and strange eloquence of the very young. In talking with me, many even reverted to the mannerisms of childhood—losing their adult composure as they related searing memories of camp life. One woman, recalling the rats scurrying across her toes at night as she tried to sleep in the hard wooden stall that was her bed, shook and shivered like a little girl. A male twin winced and whimpered like a child while telling of a painful injection Mengele had given him. But then his face lit up with a boyish grin as he described the delicious candy the doctor offered him moments after. They seemed in many ways to still be the frightened eight- or ten-year-olds of yesteryear, listening for the sound of Dr. Mengele's footsteps, dreading the arrival of the trucks that would take them to his laboratory. They wept as the memories came flooding back. Profoundly moved by their stories, I often found myself crying with them.

Only a handful of the twins were able to provide me with a complete account of their lives from the period before the war through the years after Auschwitz. Instead, most seemed obsessed by one period or another. Some vividly remembered Mengele's experiments, volunteering precise details on the blood tests, injections, X rays, and surgeries they had undergone. Others focused on the SS doctor himself—his visits to their barracks, how he liked to sit and chat with them. A few of the twins insisted they had no memories of Auschwitz whatsoever. Instead, they dwelt on the sadness of their postwar adult lives—their emotional upheavals, physical breakdowns, and longings for the dead parents they had hardly known. The younger they were at Auschwitz and the less they consciously remembered, the greater their turmoil as adults: That was the rule.

Mengele's passion for selecting Jews for the gas chambers of Auschwitz had earned him the title "the Angel of Death." With a flick of the wrist, he would consign thousands to die. Among the few exceptions were the young twins he plucked out from the selection lines for use in his research. As a genetic scientist, Mengele hoped to produce a master race of blond, blue-eyed Aryans. Twins were the key. What better way to test out theories on heredity than by experimenting on heredity's perfect genetic specimens: identical twins.

Most of the twins began their descent into Auschwitz by witnessing their entire families being led away from them to be killed. In their special barracks, located just yards away from the crematoriums, they observed the Nazis' extermination of the Jews at close range. Twins as young as five and six years of age endured torture, daily blood tests, and starvation diets, as well as facing exposure to epidemics of cholera, tuberculosis, and other deadly diseases that were rampant because of unsanitary conditions. Worst of all, of course, were Mengele's barbaric pseudoscientific experiments. But horrific as their lives were, the twins enjoyed a special privileged status, for they were regarded as "Mengele's children." And as such, they were spared the random selections and march to the gas chambers that threatened every other Auschwitz inmate.

Despite their ordeal, many of them clung to their childlike faith in life. During Mengele's mandated "recreation periods," male twins played soccer under a sky made brighter by the flames pouring out of the crematoriums. Little girls were taken to fields on the outskirts of the camp, where they picked bunches of wildflowers.

Some of the children even grew to like Mengele, substituting him for the father they had lost. "Uncle Mengele," as they called him, delighted them with candy, joked with them, hugged and kissed them. Some were persuaded that at the bottom of his evil heart was a soft spot—an untapped core of goodness—reserved especially for his twins. "I believe Josef Mengele loved little children," Vera Blau, a twin from Tel Aviv, insisted to me during our first interview. "Yes! Even though he was a murderer and a killer."

In writing this book, Mengele's special relationship with children emerged as the most puzzling—and fascinating—aspect of this "angel" of death. Monstrous as he was, Mengele still managed throughout his life to charm and beguile youngsters. The bizarre, mysterious bond forged between Mengele and "his" twins at Auschwitz remained long after they had parted company. However hard they tried, none could banish the memories of the handsome young doctor who had tortured and, they thought, loved them.

I was not alone in being touched and inspired by the twins' stories of their life under the abominable Dr. Mengele. My article "The Twins of Auschwitz Today," which appeared on the cover of *Parade* that September, was reprinted in newspapers in several countries, and prompted an outpouring of response from readers around the world. Thousands called for renewed efforts to hunt Dr. Mengele. In November of 1984, former congresswoman Elizabeth Holtzman and Nazi-hunter Beate Klarsfeld, both prominent in the battle to track down and prosecute Nazi war criminals, invited me to accompany their delegation to Paraguay as an observer in the search for Mengele. The four-member delegation, which included Bishop René Valero of the Brooklyn Catholic Diocese and Menachem Rosensaft, the child of Holocaust survivors, confronted the regime of General Alfredo Stroessner for information on the war criminal. Stroessner had once granted Mengele Paraguayan citizenship, and he was long believed to be sheltering the SS doctor. If Holtzman's mission failed to pressure the old dictator, it did for the first time awaken the United States government's interest in locating Josef Mengele.

CANDLES' dramatic weeklong pilgrimage to Auschwitz in January 1985, the twins' march across the desolate Birkenau camp a few miles away, and their emotional televised prayers at the long-standing crematorium, attracted worldwide attention. Partly as a result of their moving appeals, three separate countries decided to reopen their Mengele files. By the spring of 1985, Israel, West Germany, and the United States officially joined forces in a full-fledged hunt for the greatest killer of the Holocaust known to be still at large. A network of intelligence agencies, from the CIA and the Israeli Mossad to Interpol, agreed to cooperate to find the notorious Nazi doctor.

Mengele's children had effectively spurred the largest and most ambitious hunt ever for a war criminal. The twins quietly rejoiced at this excess of international attention, after so many years of indifference to the disappearance of their nemesis. Their experiences at the camp had made each one of them a philosopher—cynical at the ways of the world, yet strangely hopeful.

On a return trip to Israel in April 1988, I reminisced with the twins over the extraordinary events of the last four years. I found them media-savvy and, in stark contrast to our earlier meetings, quite comfortable with discussing their Auschwitz experience. Several were now lecturing around the country, and one twin, Vera Kriegel, was even flying regularly to West Germany to address audiences there about the Holocaust. Like small candles in the night, the twins were shedding light over a period of history still shrouded in strange and terrifying mystery.

My reunion with the twins also served as an affirmation of my own unique bond with them. I had been the first journalist ever to search out and systematically interview them, to help make their stories known to the world. I had rejoiced at their media triumphs and cheered their successful efforts to belatedly obtain reparations from the West German government. I felt as shocked and disappointed as they did when a body alleged to be Mengele's was discovered in Brazil in June 1985, and the global hunt was called off.

None of the twins believes the skeleton found in a lonely gravesite in Embu is that of Dr. Mengele. The weight of public opinion—and scientific testimony—has not been sufficient to convince them he is really dead. Supremely distrustful of scientists, they point to the fact that not one of the legions of forensic pathologists who examined the remains ever declared with 100 percent certainty the body was Mengele's. Since the Israeli government has consistently refused to close the case, since both Israel and West Germany continue to offer rewards for his capture, they are hopeful the real Josef Mengele will yet be found.

For them, he is still alive. They can see him standing there in his impeccable uniform, smiling, insisting he will never hurt them. How he loves to linger and play games! He enjoys seeing them dressed in the beautiful clothes he himself brings to their barracks—white pantaloons for the boys, silk dresses for the girls.

But they also cannot forget the terror of his laboratory, the blood tests, the injections, the experiments, the murderous operations.

Mengele's twins are now inching into old age. Most are married, with children and grandchildren of their own. But despite the presence of spouses, friends, and neighbors, they have periods of despair, when the present seems far less real than the past. When no one is looking, they pull out faded letters and old photographs. With trembling hands, they peer at pictures of dead parents and siblings, or of themselves before the horror. These are the times they realize they will never escape the shadow of the Angel of Death. While going about their daily chores, they may catch a glimpse of his face with its cynical half-smile, the sadistic glint in his eye. They see the legions of skeletal men and women being carted away to their deaths. And then they remember the other children, their arms and legs flecked with bruises from Mengele's needles: the twins who, unlike themselves, did not survive.

Many of the twins suffer from a recurrent nightmare that they are back at Auschwitz. In their dreams, angry dogs bark while tall men in uniforms relentlessly pursue them across dark, ominous terrains. And when these looming figures catch up with them, as they invariably do, Mengele's twins cry in their sleep, as they surely cried as children so many years ago.

Their nights are also filled with the ghosts of their loved ones. Again and again, they relive the moment their parents, brothers, and sisters were herded into the gas chambers. They hear their family's resigned footsteps trudging into the stark rooms disguised as showers and imagine their surprise and terror when gas gushes out instead of water.

Hedvah and Leah Stern find that as they grow older, they think about their mother even more than before. They can clearly see the lovely young woman in a black print dress, waving good-bye as she promises they will be reunited. In the stillness of an Israeli night, Hedvah and Leah hear their mother shouting to them, "Wait, children, wait—wait for me at the gate." Her anguished cries have carried over the busy, eventful years, like an echo reverberating louder and louder over a deepening abyss.

For most, the passage of time has failed to bring peace of mind. Nearly half a century later, they feel guilty at having been born twins. After all, despite all they suffered as Mengele's guinea pigs, they still enjoyed the privileged status he bestowed upon them. Mengele's twins are condemned to live with a terrible double-edged sword: the hell they endured because of Mengele, and the life they owe him as "his" children.

"Why were we spared?" Mengele's twins ask themselves today. "Why were we the only ones in our family to survive?" The more fortunate ones turn to their lone surviving sibling for comfort and answers, only to find the same sense of bewilderment and despair.

The sights of the camp—the dead bodies, the bones, the crematoriums spewing bright red flames into the vermillion skies—have continued to haunt Mengele's twins. They emerged from the ashes of the Holocaust dazed, anxious to leave the past behind them, but they have learned the past will never leave them. And although they last saw Josef Mengele in the winter of 1944, he has remained an ever-present force in each of their lives. This is his story—and theirs.

—LUCETTE MATALON LAGNADO

I came to the story of Mengele's children through my late husband, and although I am an American, his story became mine. I met Alex Shlomo Dekel while living in Israel in January 1962, and we were married one year later. My American upbringing had intrigued him; his European past had fascinated me: the gallant widowed mother, who had raised him and his brother before they were transported to Auschwitz in the first convoy from his native Hungary; his experiences in the concentration camp as a Mengele guinea pig; and his liberation from Birkenau by American GIs, who saved his life at the time by intravenous feedings.

Alex was not a twin, but he had been spared an almost certain death because of his Aryan good looks and his German-language fluency. Not long after we started seeing each other, he began to tell me his story. Especially haunting, I thought, was his final memory of his mother, whom he had spotted despite her shaved head in an adjacent yard behind a barbed-wire fence. As one of Mengele's "privileged" children, he knew he had more rations than she, and so he saved up all his food for the day to give to her. She refused to take it from her child, so he threw it over the fence, where she was forced to retrieve it. Later he told me that he had survived by dreaming at night of consuming huge meals, and that this had given him the strength to face the starvation of the next day. Nourished on dreams!

That he had been a child prisoner of Auschwitz I understood. That it would color his entire life I did not. But after we were married, I discovered and gradually came to share his all-consuming passion to find Mengele and bring him to justice. Auschwitz was never far beneath the surface of his thoughts. He would try to forget, but from time to time it would emerge, in episodes of unexplained rage and in periods of deep depressions that I now understand are scorched into the Auschwitz legacy of all the survivors.

Israel became his raison d'être. I told him it was his revenge. His life was often in danger, starting with his swimming ashore illegally from Cyprus where his ship had been detained by the British, through his dare-devil exploits, first in the Haganah and, later, in the service of the Israeli government (some of which still cannot be revealed). His entire life was shaped by forces outside of himself. Like so many others, Israel became his lost family. I sensed he was sometimes more at home in the memories of Auschwitz than in the real world. For the survivor, all is always Auschwitz.

As an archivist attached to the Israeli consulate in New York, Alex began to collect information on Mengele's every move. At first, I didn't pay much attention to the pieces of information he shared with me on the death-camp doctor's whereabouts. Given enough time, I was sure I could make him forget. After all, we were living in America now. We had each other. But being married to him was like being married to a secret agent. There were always secrets. His wasn't exactly the kind of work a wife could innocently ask about at the end of the day. "What did you do today, dear?" I tried that only once!

His compulsive search for Mengele gradually escalated into a full-scale, one-man hunt. He began to contact anyone who had ever heard of Mengele and might have knowledge of him. I remember being awakened one night by a telephone call at 3:00 A.M. from South America: Someone had phoned to report a sighting. There were many such calls. The papers Alex accumulated began to take on the dimensions of a personal library. Every newspaper clipping and magazine article, anything related to Mengele's whereabouts, helped Alex put pieces of evidence together in the manner he had learned during his twenty-five years as an intelligence officer for the Israeli government. Alex soon pinpointed Mengele's various residences and many of his movements in and out of South America. He even located a wholesale pharmacist supplying drugs to Brazil—and concocted a wild scheme to take Mengele's medicine to him (which I vetoed as being far too dangerous). Anyone who had even the remotest connection with Mengele or with South America was a welcome visitor in our home.

In 1976, Alex contacted the editors of *Time* and convinced them of the critical need to locate Mengele. They financed his research as he pursued the trail all over Europe, uncovering important new facts. From Vienna, where he conferred with Simon Wiesenthal, he went to Poland, where he persuaded the government to open its files on Auschwitz and Mengele's experiments there. Elated, he brought the evidence home, where I slept innocently with Josef Mengele's fingerprints and SS files under our bed.

Time published these findings in September 1977, and soon after, Alex began to set down his experiences for a memoir that he hoped would someday help arouse the world's conscience and lead to Mengele's arrest. In the meantime, he initiated another Mengele story with *Life* magazine and spent hours with Dr. Robert Lifton, giving oral testimony for Dr. Lifton's book on the Nazi doctors. And although at the beginning I had little patience with Alex's hunt to find Mengele, I could no longer ignore the major theme of his life. "Live now," I had urged him. "Remove that tattooed reminder from your arm," I begged. "Remove my arm," was Alex's response. Would I have married him if I had known how his life would be consumed by this obsession? Yes, I know I would have.

That fateful summer of 1983, we were planning to go to Israel, where a high-ranking intelligence contact had promised Alex access to Israel's file on Mengele. Before that, the man arrived in New York in June, and when Alex returned home after meeting his plane, I noticed he was distracted, talking of many things but not about Mengele. Nor was he feeling well. He brushed aside my pleadings to go to the hospital when he complained of pain in his left arm. His lack of concern for his own well-being was irresponsible but typical. He always felt immune from the ravages of time and fate—after Auschwitz, what else could happen to him?

When Alex died suddenly after suffering a massive stroke and heart attack, I realized retrospectively what his contact must have told him that night; what the Israeli government already knew: Josef Mengele was dead. I could not help but feel that the hunted had finally caught up with the hunter. This knowledge surely must have broken Alex's heart, dissolved his resolve to live.

After Alex's death, I resolved to carry on his mission. I met Lucette Lagnado, whose own interest in Mengele and his child victims paralleled Alex's. Many events took place in the course of writing this book, including the discovery of Mengele's remains on June 21, 1985—two years to the day after Alex's funeral. Their destinies would remain inextricably intertwined in death as they had been in life.

"I am sad and heartbroken," Elie Wiesel had telegraphed me from Washington when he heard of Alex's passing. "He was one of our most devoted companions. Few have endured his agony. Few have lived with it as deeply. For myself and all the members of the United States Holocaust Council, we shall always remember him." As I sorted through Alex's papers to gather material for this book, I came across this fragment repeated throughout his notes:

Tell your children of it,
And let your children tell their children,
And their children another generation.
—*Joel 1:3*

It is this spirit that has guided me in completing his work.

—Sheila Cohn Dekel
New York City

CONTENTS

DRAMATIS PERSONAE

TWINS' FATHER: Born 1915 in Budapest, Hungary. Real name: Zvi Spiegel. Family moved to Munkaks, Czechoslovakia. Was twenty-nine years old when he and twin sister, Magda, were deported to Auschwitz. Assigned by Dr. Mengele to be in charge of the twin boys. The children called him *"Zvilingefater"* or "Twins' Father."

MAGDA SPIEGEL: Twin sister of Zvi Spiegel. Married with a seven-year-old son when deported to death camp. Worked as Mengele's cleaning woman.

HEDVAH AND LEAH STERN: Identical twins. Born in Hungary in 1931. Age thirteen and a half when sent to Auschwitz with their mother. The Stern sisters, as they became known in the camp, are virtually indistinguishable from one another. They speak in one voice, agree completely with what the other says, feel what the other feels, think as the other thinks.

MOSHE OFFER: Born 1932 in small town in Hungary (now a part of Soviet Union). Deported to Auschwitz at age twelve, along with his twin brother, Tibi, parents, and five brothers. Entire family, except for twins, sent immediately to the gas chambers.

ZVI THE SAILOR: Born Zvi Klein in Galicia. Not quite thirteen years old when he and identical twin brother, Ladislav, were deported to Auschwitz, along with extended family. All but the twins perished. Brothers estranged since the war.

EVA MOZES: American founder of CANDLES, the international organization of Mengele twins. Launched drive in 1984 to find and reunite all the twins and publicize their story. Born in Cluj, Romania, in 1935. Deported to Auschwitz in the spring of 1944 along with her identical twin sister, Miriam, her parents, and two older sisters. Entire family slaughtered except for twins. Currently resides in Terre Haute, Indiana.

MIRIAM MOZES: Twin sister of Eva, now living in Israel. Co-founder and organizer of CANDLES in Israel and Europe. Works as head nurse in Israeli hospital. Suffers severe health problems believed due to experiments undergone at the hands of Dr. Mengele at Auschwitz.

JUDITH YAGUDAH: Born May 25, 1934, in Braşov, a small town in what used to be Hungary, but is now part of Romania. Nazis sent entire family to ghetto in Cluj, and from there to Auschwitz. Was ten years old when she, identical twin sister, Ruthie, and parents were deported to Auschwitz. Father killed immediately by Nazis. Judith and Ruthie sent with their mother to twins' barracks.

OLGA GROSSMAN: Born in 1938 in Czechoslovakia. Was six years old when deported to Auschwitz with her twin sister, Vera. Family spent time in other concentration camps before Auschwitz. Has absolutely no memories of her stay in the camp, and is unable to talk about the war. Suffers from recurrent hallucinations about Mengele. Frequently institutionalized as an adult.

VERA GROSSMAN: Olga's fraternal twin sister. Was so different in appearance from her sister that Mengele did not believe they were really twins. Unlike twin, claims to have very vivid memories of Auschwitz. Has no difficulty discussing the war. Outgoing. Active in CANDLES and other Holocaust groups. Travels to Germany to lecture about the twins of Auschwitz. Served as author's interpreter for all Hebrew-language interviews of Israeli twins.

ALEX DEKEL: Born in Cluj, Romania, in 1930. Age thirteen at time of deportation to Auschwitz. Though not a twin, selected by Mengele to go to the twins' barracks because of his striking "Aryan" features. Subjected to medical experiments like the twins.

PETER SOMOGYI: Born in Pécs, Hungary, in 1930. Deported to Auschwitz in the summer of 1944 on one of the last transports of Hungarian Jews. Mother and sister perished in the gas chambers. Only Peter and twin brother survived. Extremely popular with Dr. Mengele—he nicknamed them "the intelligentsia" because of their fluency in several languages, their knowledge of classical music, and their ability to play the piano.

VERA BLAU: Was eleven years old when deported to Auschwitz in April 1944 from Czechoslovakia. Arrived with twin sister, Rachel, mother, and little brother. Only twins survived. Vera, an artist in Tel Aviv, insists Mengele "loved" little children.

MENASHE LORINCZI: Born in small town in Romania in 1934. Shortly after Menashe and twin sister, Lea, celebrated their tenth birthday in the Cluj Ghetto, they and their grandparents were rounded up and deported to Auschwitz. Grandparents immediately killed. Fate of mother unknown. Became a messenger boy for Mengele. One of the few child inmates permitted to wander freely around the death camp.

LEA LORINCZI: Twin sister of Menashe. Became an active Communist in Romania after World War II. Later emigrated to Israel, married ultra-Orthodox man, and joined Hassidic sect. Moved to Williamsburg, Brooklyn. Now owns ladies' garment store on Manhattan's Lower East Side.

EVA KUPAS: Deported to Auschwitz with twin brother in spring 1944. Birth date unknown. No recollection of prewar life. Her only memory of the death camp is the day she and other twin girls were taken to pick wildflowers in a field just outside of Birkenau.

SOLOMON MALIK: Deported to Auschwitz in May 1944, at the age of thirteen, along with parents, twin sister, and another pair of twin brothers. Only the four twins survived.

JOSEF MENGELE: Born March 16, 1911, in Günzburg, Germany. Son of wealthy factory owner. Became a scientist with a special interest in twins. Volunteered to go to Auschwitz to work as a doctor in the spring of 1943, at age thirty-two. Disappeared after the war. Also known as Helmut Gregor, G. Helmuth, Fritz Ulmann, Fritz Hollmann, José Mengele, Peter Hochbicler, Ernst Sebastian Alvez, José Aspiazi, Lars Ballstroem, Friedrich Edler von Breitenbach, Fritz Fischer, Karl Geuske, Ludwig Gregor, Stanislaus Prosky, Fausto Rindon, Fausto Rondon, Gregor Schklastro, Heinz Stobert, Dr. Henrique Wollman, and "Beppo."

I remember picking flowers at Auschwitz. One day a warden took me and a group of twin girls to a nearby field. We picked all sorts of wild flowers. Then we walked around the camp holding bunches of flowers for everyone to see.

And as we were walked through the camp, women prisoners started shouting out names—this name, that name.

They were mothers, you see, and they were looking for their daughters as we passed by.

I heard them cry out, "Maybe my child is among them."

—EVA KUPAS, twin at Auschwitz

A couple of years after the war, our rabbi in Cluj decided to hold a memorial service for all the Jews from our town who had died in the Holocaust.

The rabbi said that if anyone in the congregation had bars of soap left over from the concentration camp, we should bring it to the temple. He told us it had to be "buried" because it had been made from human flesh.

It was the first time I had heard that. Before we left the camp, my twin sister and I took whatever we could with us. At a time when goods were scarce, I used the soap all the time.

After the rabbi's address, I felt terrified. I thought, "Maybe I used soap made from my family."

For years, I had continuous nightmares. Every night, I dreamt I was washing myself with soap made from my parents or my sister.

—EVA MOZES, twin at Auschwitz

Prologue

THE JAZZ BAR

ZVI THE SAILOR:

All my life, I have tried to run away from Auschwitz. I have sailed everywhere, from the North Sea to the Tropic of Cancer, the Great Lakes of Michigan to Madagascar.

Montevideo, Johannesburg, Hollywood . . .

But wherever I went, thoughts of the death camp came back to haunt me.

I try to forget. I try to erase the memories. But I can't seem to leave Auschwitz behind, no matter how far I go. I find reminders of the camp all over the world. Even right here in Israel.

From the window of my apartment house in Ashdod, I can see a factory—a large industrial plant that manufactures I don't know what. Textiles, I think. It has a large chimney which spews out fire at night.

And when I look out my window in the evening, I see not a factory chimney, but the crematorium of Birkenau with its flames, its tall red flames leaping out at the sky.

Last night was Lag b'Omer, the feast of the mystics. The festival of bonfires. Once a year, we Jews are supposed to celebrate the end of the mourning period for our lost sages.

All over Israel yesterday, the little children were celebrating by building bonfires. As I walked along the streets, I watched them joyfully throwing bits and pieces of wood into the flames. They roasted potatoes, sang songs, danced around the fire.

Lag B'Omer is only once a year, but I am always seeing fire. Day and night. All year long.

As I watched the children making their bonfires yesterday, I thought of the fire made OF the children. The fire without wood. The fire of Auschwitz-Birkenau.

For four months, forty years ago, I also saw fire day and night. I lived in barracks with a group of twins. Twins! Twins! Dr. Mengele only wanted twins.

Our barracks were located just a few yards away from the crematorium. We could see it from our window, from our door, wherever we turned we saw fire. Every day and every night those four months, we watched as Mengele motioned people toward this crematorium.

We saw thousands of people going in each day. And not one ever came out.

We were so close to this crematorium that we became friends with the Sonderkommandos, the young boys who worked inside the gas chambers. A few of them were from my hometown in Galicia.

They gave us detailed reports: "Today, we burned ten thousand Jews. Yesterday, we disposed of eight thousand."

And every day, they warned us, "You will be next. Dr. Mengele has decided you should be reunited with your parents." We believed them. We were sure that tomorrow, maybe the day after tomorrow, Mengele would send all the twins into the flames.

But even as we expected to die, we continued to obey Mengele. He made us write cheerful postcards: Everything is fine, we wrote to other Jews. We are working.

I see fire everywhere. Although I am at sea, I am consumed by the flames.

Yes, yes, there have been moments when I forgot.

I sailed to the United States a few years ago. My boat docked in New York City. I had a few hours off.

First, I visited an aunt in Brooklyn. But I felt restless and uncomfortable. She kept talking and talking, and my mind was elsewhere.

In Auschwitz, I suppose. I finally told her my ship was sailing, and left her house.

I started walking around Manhattan. I was wandering on the West Side, along Eighth or Ninth Avenue, when I saw a large crowd of people standing outside a bar. They were pushing their way in, and I found myself being pushed inside along with them.

Inside, it was dark and cool. I could see musicians lined up on a stage. There were forty, fifty, of them and they were standing in a single straight line. And even though they had no leader, no conductor, they managed to play in perfect harmony.

They were playing music that was not from this earth.

Among the musicians, I noticed "Satchmo"—Louis Armstrong. I recognized him instantly from photographs.

I ordered a beer. And another. Then another, and another. It was very expensive; you had to leave a tip after each order.

At a table not far from me, there were these two young girls. They sat there, laughing and drinking.

Then, they got up and went over to another table where a couple of young men were sitting, and started kissing them. They embraced to the rhythm of the jazz.

I was in that bar for an hour, the only white man there. I spent sixty dollars—more money than I've ever spent in one hour.

Before I even realized what was happening, the same crowd that had pushed itself in, that had pushed me in, got up and left. I was pulled outside with them.

I found myself back on the street. I started walking, and somehow, even though I was quite tipsy, I made it back to my ship.

I have been to New York many, many times since then, and always I have searched for that jazz bar. But I have never been able to find even a trace of it.

I can still remember the music, the atmosphere, the beer, the girls kissing the boys to the beat of Louis Armstrong.

Yes, I was happy then! In fact, never in my life have I been as happy as during the hour I spent inside that dive on the West Side.

Because for one whole hour, I actually managed to forget Auschwitz, and Dr. Josef Mengele. . . .

1

MENGELE AND HIS CHILDREN

Nine-year-old Beppo Mengele stood at the Günzburg railroad station, awaiting the train that would bring in supplies for his father's factory. How excited he became when it pulled into the station! The townspeople recall how, in a loud voice, he would order his two brothers to get ready to unload. Young Mengele was always happy when the transports arrived.

Karl and Walburga Mengele often assigned their oldest boy the task of overseeing the shipment of goods to and from the Mengele farm-equipment factory. Beppo would ride the horse-drawn wagon down to the train station, delighting at the ruckus it caused as it clattered across the cobbled stone streets of the sleepy little town. If it was early in the morning, the residents of Augsburgstrasse, Günzburg's main street, were awakened by the noise of the heavy steel parts banging and clanging against one another. They would sigh and mutter affectionately, "Mengele is coming." With his dark hair, gleaming eyes, and mischievous grin, Beppo was an endearing child—the most popular of the three Mengele brothers.

* * *

At the station, the little boy proved that he deserved the trust his parents placed in him. He watched carefully as the shipments were stacked onto the trains. If a train had brought in supplies, he supervised their loading into the carriage, making certain nothing was broken or left behind.

It was a grown-up job, and young Mengele was said to revel in it. A quarter of a century later, an older Beppo—SS Dr. Josef Mengele—still delighted at the arrival of trains and their cargoes, but at a different railway stop.

MOSHE OFFER:

I was born in 1932 in a small town in what is now a part of Russia. My twin brother and I were the youngest of four brothers. We were nicknamed "Miki and Tibi."

My family was very wealthy. Father owned a large estate with two plants that manufactured liquor. We grew potatoes and fruit on our farm, and made sweet liqueurs from them.

It was a wonderful life. Each morning, Tibi and I were taken to school by a horse-drawn carriage. We owned two horses!

Then the Germans came in 1944. They took everything from us. They confiscated our gold, our jewelry, our furniture—all our possessions. We were no longer permitted to live on the farm. Instead, we were ordered to take two suitcases and report to the village synagogue, where we were jammed along with all the other Jews in our town.

From there, we were moved to a ghetto in a big city. The conditions were terrible. We had no clothes. We were forced to wear—literally—the shirt on our back. We weren't given food, either, and we were constantly hungry. Dead people littered the streets; they had starved to death.

One day, the Germans returned. We were taken out of the ghetto and placed in cattle cars. The journey took eight days—eight days without water, without food. It is painful for me to remember what went on there. It is too horrible to describe.

We arrived at Auschwitz in May 1944. I can even tell you the exact time: ten o'clock in the morning.

When they opened the doors to our cattle cars, there were a lot of dead children. During the trip, some mothers couldn't bear to hear the cries of their hungry babies—and so they killed them. I remember two blond, very beautiful children in my car whose mother had choked them to death because she could not stand to watch them suffer.

When we stepped off the trains, we could hear soldiers yelling, "Men on one side, women on the other side." Some German SS guards were also shouting, "We want twins—bring us the twins!"

Dr. Mengele was making the selections. He stood there, tall, nice-looking, and he was dressed very well, as if he wanted to make a good impression.

He had very soft hands, and he made fast decisions.

I heard my father cry out to them he had twins. He went over personally to Dr. Mengele, and told him, "I have a pair of twin boys." Mengele sent some SS guards over to us. My twin, Tibi, and I were ordered to leave our parents and brothers and follow them.

But we didn't want to be separated from our mother, and so the Nazis separated us by force. My father begged Mengele to give us some food and water. But Mengele motioned to an SS guard, who beat him up on the spot.

As we were led away, I saw my father fall to the ground.

There is nothing in Josef Mengele's early life that would have prepared him for the notoriety that was destined to engulf him. As a youth, he was charming and carefree and not especially studious. No one who remembers Mengele growing up in his picturesque Bavarian town ever saw a hint of the pathology that would make him a killer, or a sign of the obsessions that would make him a concentration-camp doctor. There was an innocence and a sweetness to young Josef that would lead Günzburg's citizens to shake their heads in disbelief when they heard, years later, of his savage deeds at Auschwitz. The fiendish death-camp physician had nothing in common with the lovable youngster they all had known. The Nazi professor brutally experimenting on young twins could hardly have been the same playful little scamp they affectionately called "the Beppo," years after he had grown out of the childish nickname.

Even as he grandly swept through the barracks at Auschwitz years later, he was like a vision, this handsome, genteel German officer in his impeccable SS uniform, shiny boots, and white gloves. He looked less like a Nazi official than a Hollywood version of one—Tyrone Power in the role of SS captain. Dr. Josef Mengele would maintain this beautiful facade throughout his tenure at Auschwitz. None of the bewildered new arrivals would discern the murderer, or even the sadist, in the polite young SS doctor until it was too late. Mengele would decide who lived and who died with a smile and an airy wave of his elegant white-gloved hand. He would charm the women of Auschwitz-Birkenau even as he sent them to the gas chambers. The Gypsies would love him as one of their own to the very end. But Mengele would be at his best with the young twins he removed from the selection line for use in his medical experiments. With them, he could be as warm and affectionate as he had been as a little boy growing up in a small Bavarian farming town. Did he see something of his old self in these children, innocent and doomed? For what better symbol, after all, of Mengele's own dual nature—the angel and the monster, the gentle young doctor and the sadistic killer—than a twin.

Josef Mengele was born on March 16, 1911, in Günzburg, a medieval village on the banks of the Danube. Three years earlier, his mother, Walburga, had given birth to a stillborn child. Soon after, she became pregnant again. Within sixteen months, Josef had a brother, Karl Thaddeus. In 1914, another son, Alois, was born. Josef adored "Lolo." He felt much closer to him than to Karl Jr., and, as they grew up, always included him in all their games.

In the years before World War I, the Mengeles lived modestly, sharing a house with another family. While Walburga tended to the boys, Karl Mengele spent most of his time expanding his new factory. Karl was new to Günzburg. He had left his own native village after his older brother inherited the family farm, and studied to become an engineer. Settling in Günzburg, he had married the strong, energetic Walburga, four years his senior.

Walburga's parents, who were wealthy farmers, loaned their son-in-law the money to start his new business. Karl quickly proved he deserved the investment by patenting a number of handy farming tools, which enhanced his standing in his wife's hometown. What precisely these tools were has been lost in Günzburg's history—but their inventor, though long dead, still enjoys a reputation as a creative, industrious man.

World War I was the turning point for the Mengeles' fortunes. The family received lucrative contracts to manufacture military goods and was busy turning out army vehicles and other weaponry. At the end of the war, once again in the farm-equipment business, the Mengele plant was among the largest in Günzburg. By 1918, the Mengeles were rich enough to move to a home of their own. Karl and his wife picked a lovely villa across the street from the Gymnasium the boys would attend.

VERA GROSSMAN:

I was born in Czechoslovakia in 1938 to a very wealthy family. My father, who was twenty years older than my mother, owned many fields and plantations. More than two hundred laborers worked for him, tilling the soil and helping to pick and package the fruits and vegetables.

He fell in love with my mother when she was just a young girl. Before the war, Mother was extremely beautiful. She had jet-black hair and blue eyes.

When they married, he brought her to live on his estate and lavished her with clothes, servants, jewels. All her dresses were hand-made in Prague.

Once, a dress she had ordered for a wedding didn't arrive on time. My father sent his chauffeur all the way to Prague—hundreds of kilometers away—to fetch it.

My father was delighted when Olga and I were born. At his age, he considered twins a double blessing. We were spoiled and given everything children could want.

Of course, this lifestyle came to an end when the Germans invaded Czechoslovakia. Jews could no longer live openly. Even though my father was very rich, and had connections with the government, we still had to go into hiding. We had to abandon everything.

First, my father bribed a Christian family to let us live in their attic. It was a terrifying period. I was only four years old, but I remember how we were constantly admonished not to make any noise. Even to cry was forbidden, because it would endanger the family.

And to this day, I am haunted by this feeling—that if I do something wrong, my whole family will die.

Eventually, it became so dangerous to hide Jews in Czechoslovakia that no one wanted to shelter us—no matter what my father was willing to pay. We lived like animals, lying low during the day, foraging for food at night. I can recall eating raw potatoes, when that was all my parents could find.

But eventually the Germans found us there. We were sent to a series of concentration camps, until we arrived finally at Auschwitz, in the spring of 1944.

To outsiders, the Mengeles seemed a close, devoted family. A devout Catholic, Walburga raised her sons to be regular churchgoers. Old family albums show Josef dressed as an altar boy, the picture of innocence and piety. Dressed in their Sunday best, father, mother, and the three children went each week to the beautiful eighteenth-century church near the old marketplace. They were a handsome family and, as their wealth increased, the cause of some fascination in the town. Neighbors recall how Josef, Karl Heinz, and Lolo shared friends, romped about the fields, and went on frequent outings with their parents. A few older residents can still remember the skating parties Karl and Walburga held for the children on a small pond near their house. They served delicious candied apples, while music from a wind-up gramophone played in the frosty air.

The Mengele boys were always the object of much fawning, especially Josef. He was a docile child, and eager to please; Walburga had made sure of that. But though he acted like an angel, he looked more like a young Gypsy. In fact, some who knew him then had the distinct feeling that at any moment this obedient little boy would make a run for it, defying his parents as other normal children did.

Mengele never broke loose, however. His early school records show he was a model child, who impressed his teachers with his exceptionally good behavior. Though a mediocre student, he still managed to receive A's in conduct and diligence throughout his elementary and high school years. Even in the strict Prussian atmosphere of a prewar Gymnasium, teachers went out of their way to praise the perfect conduct of Beppo Mengele.

Alex Dekel:

I could hear the blaring music of Lohengrin being piped through loudspeakers as I walked through the gates of Auschwitz. It was like entering the inferno.

I was thirteen years old when my mother and I were deported to the death camp from our hometown of Cluj, in Transylvania.

In early March 1944, my mother had received word that we were being sent to a work camp in central Hungary, supposedly to help with the war effort.

We were afraid, but the hope of living, of going only to a labor camp, kept us going.

The deportations were organized alphabetically, and since our last initial was D *we were the first to be called to board the train.*

After two horrible days aboard this train, I knew we had gone far beyond the borders of Hungary, and were destined either for Germany or Poland. Panic reigned in the cars. Two people committed suicide.

My mother clutched me to her and covered my ears with her hands.

When the train finally stopped, the Germans ordered everyone to get out. I smelled a faint burning odor. A sign along the tracks read BIRKENAU.

Dr. Mengele was standing at the head of the selection line. He noticed me immediately because I didn't look Jewish. I had very blond hair, and blue eyes, and I was in excellent physical shape. When he started talking to me, I answered him in fluent German.

Mengele wasn't only looking for twins—he wanted triplets, midgets, hunchbacks, any unusual types. Even people like me—Jews who looked like perfect Aryans. He asked me to step out of the line. I looked around for my mother, but she had disappeared.

I prayed that she was among a group of women who had been selected to live.

Mengele's mother was the archetypal German hausfrau whose life revolved around her children. Photographs show a heavyset woman with a stern, homely face and dark, scowling eyes. Unlike her husband, who adopted an aristocratic demeanor with his growing wealth, Walburga made no attempt to alter her dowdy and matronly appearance. In the fashion of older peasant women, she dressed almost entirely in black.

"Wally" Mengele suffered from a terrible weight problem, which stemmed from her lone indulgence: food. She simply loved to eat. One Günzburg woman who knew her well, now relocated to New York's affluent Westchester County, spins story after story about this impassioned craving. Mrs. Mengele's favorite pastime, she recalls, was the afternoon kaffee klatsch, the get-togethers with women friends over coffee and pastry. Over the years, Walburga grew enormous. She ate constantly, compulsively. She became so obese, she could hardly walk. She was so massive, she looked almost pregnant. She was so hungry she devoured everything in sight.

There was a troubling, terrifying side to Mengele's mother that only a few people saw, such as the workers at her husband's factory. Dr. Zdenek Zofka, the unofficial historian of Günzburg, says the employees fretted whenever Wally came to visit. She had no compunctions about yelling at them and embarrassing them before the others, according to Zofka, and was inclined to fly into rages at the slightest provocation. They nicknamed her "the Matador," and instinctively stayed out of her way. Once, she screamed at some female employees for not having washed the factory's curtains. When they argued it was not their job, she continued to scold and threaten them, thereby earning their lifelong enmity. Over forty years after her death, old Mengele factory workers still harbor bitter feelings toward the indomitable Walburga Mengele.

Mengele's mother was larger than life, and she loomed as a gigantic figure in Josef's life—impossible to escape, equally impossible to please. She could be warm and maternal, or she could behave like a raging bull. Her reactions were impossible to predict. In an unpublished autobiography he wrote many years after the war, Mengele recounted a day when his father came home with a wonderful surprise for the family: a new automobile. The three boys were overjoyed. Karl invited his wife to come out and join them for a ride. But Walburga was livid. How dare he indulge in such a large purchase—such an extravagance—without her approval? Karl tried to soothe her, to no avail, and finally exploded and threatened to leave her. According to his account, Josef listened, petrified, to his parents' quarrel. After his father had left the room, the little boy went over to comfort his mother. "I will always stay with you," he told her.

HEDVAH AND LEAH STERN:

When they opened the door to our cattle car, our mother became very frightened. "Stay with me, children," she told us, refusing to let go of our hands.

But then some prisoners told her in Yiddish, "Tell them you have twins. There is a Dr. Mengele here who wants twins. Only twins are being kept alive."

But our mother didn't want to be separated from us. She said,

"No, you are coming with me," and continued walking toward the crematorium.

We were thirteen-and-a-half years old when our family was sent to Auschwitz from our small town in Hungary. There were a lot of Jews living in that town before the war.

Our mother was a widow. Father died when we were only six years old, and mother never remarried. She decided to raise us by herself because she feared that a second husband might mistreat us.

We were very close to our mother. She was a seamstress, and after our father died, she had to work hard to support us. Yet she never let us feel like orphans. She gave us whatever we wanted. We lacked for nothing.

We adored her. We fought with each other to get her attention. We dreamt of the day we would be old enough to work and help her.

We hoped to open up our own seamstress shop. Whatever we earned, we planned to give it to our mother. We wanted to call it "The Stern Sisters."

At Auschwitz, mother was determined to hold on to us. She hid us under her skirt.

Josef was the one person who was able to pierce his mother's stern facade and elicit a smile and a little warmth. Mengele's writings about his childhood suggest that from the start, he was his mother's son, loyally siding with her, whatever the issue. He seems to have preferred his temperamental mother to his father, whose life revolved around the factory. Although Mengele patterned himself in dress and demeanor after Karl, his love was reserved for Walburga. Twice married, he never lived with any woman for any length of time. Also twice separated, he had relationships marred by estrangements and disaffection. He could not form a deep attachment, even to the mother of his child.

When he was fifteen years old, Mengele became ill with osteomylitis, a disease of the bone marrow that was almost always fatal in those days. He also developed two ailments that were probably related to the osteomylitis: nephritis, a painful inflammation of the kidneys, and a severe systemic infection. Because of his illness, he was out of school between six and nine months. His grades, never very good, suffered a dramatic decline. It was a struggle simply to keep up with his classmates and not be left back. He failed several subjects, prompting one teacher to observe on his report card that Josef "must become more diligent, more studious, and more ambitious."

Outside the confines of the dreary Gymnasium, Mengele fared much better. His social skills were far more impressive than his intellectual prowess. To his friends and schoolmates, young Josef was charming and articulate—a natural leader. The handsome, debonair Beppo was in his element in Günzburg's chic café society and salons. Mengele and his youngest brother, Lolo, ran with an elite, affluent crowd of young men and girls who were pretty and flirtatious without being loose. Josef and his friends dressed impeccably, soon drove their own cars, and never left home without the requisite hat and white gloves. They didn't actually wear the gloves, but only held them, nonchalantly, in one hand.

It is ironic that years later, in his autobiography, Mengele conjured a picture of himself as a solitary, self-effacing youth. In his own eyes, he was an exceptionally virtuous young man, serious and committed, with few interests other than his studies and political youth groups. He depicted himself as monklike and ascetic, different from other young men his age.

But Mengele's vision of his past was, in fact, an excellent description of his lonely existence during the period he penned his manuscript, when he was already in his sixties. He was merely projecting what his mother had wanted him to be—austere, chaste, self-denying—whereas his actual nature tended toward luxury and self-indulgence. Like his father, he preferred the finer things in life.

At balls, which were held throughout the year at the homes of his wealthy friends, Beppo Mengele was one of the most sought-after dance partners. He was extremely handsome, with classic features and a slight enigmatic smile. Most striking were his eyes, which were wide and thoughtful and varicolored. Josef was certainly conscious of his good looks, though he wished he were taller. Like his father, he took great pains to appear the picture of elegance. His suits were always of the most expensive fabrics, tailored in the most flattering cut. Yet acquaintances remember he managed to seem as though he hardly spent a moment on his appearance. Young Mengele possessed a grace, mingled with nonchalance, which the Italian Renaissance princes had dubbed *sprezzatura*—the art of going through life looking as if whatever you did came effortlessly.

SOLOMON MALIK:

Dr. Mengele made an immediate impression. He was very handsome, very nicely dressed.

When my family arrived at Auschwitz in May 1944 from Romania, we heard Mengele asking for twins. There were two sets of twins in the family—my sister and I and our two younger brothers. But we didn't want to admit we were all twins. We did not know what was going to happen.

Then, another family who knew us pointed us out to Mengele, telling him, "There are some twins!" And so Mengele took the four of us away, along with our mother.

❧

EVA KUPAS:

He was so nice-looking. . . .

We arrived in Auschwitz in the spring of 1944.

My twin brother and I were marching with our mother toward the gas chambers when she told us, "Children—go to the Germans. Run back to where they are taking the twins."

Her instinct, I guess, told her we would be safer if we left her. While our mother and little brother continued marching in the direction of the crematorium, we ran back toward the selection lines.

There was Dr. Mengele, standing at the head of the line.

My first impression of Mengele was that he was so nice-looking—very nice-looking.

Young Beppo's elite group of friends adopted a favorite hangout, the fashionable Cafe Mader in the center of town. They spent most Saturday nights there, whiling away the late hours, sipping coffee, and chatting, thoroughly indifferent to the blandishments of a rabble-rouser named Adolph Hitler whose influence was just beginning to be felt in the late 1920s. Mengele's crowd was completely uninvolved with Günzburg's burgeoning Nazi circle, caring much more about expensive cars than politics.

But Mengele and many of his friends did become active in a group known as the *Grossdeutscher Jugendbund*, or Greater Germany Youth Movement. The young people's branch of the *Stalhelm* or "Steel Helmets" party, it was an intensely nationalistic and patriotic organization that urged its young members to strive for a return to the *Deutschland* of old, the *Volk*—basic values of the earth and "Mother Germany." In some ways, the group resembled the back-to-the-land groups such as the kibbutzim of Palestine. The old veterans of World War I who formed the original Steel Helmets recruited many young people to perpetuate their goals. Josef became a local leader, preparing papers and giving eloquent speeches.

According to Zofka, the author of a doctoral dissertation on the political allegiances of Günzburg's population before the war, dozens of young men, many of them sons of the town's elite, also gravitated toward this movement. The upper-middle-class German youth of that era volunteered for the *Grossdeutscher Jugendbund* almost as reflexively as their American counterparts were signing up for the Boy Scouts. Mengele's strong allegiance to the *Volk* movement continued past his adolescence. He officially joined the *Stalhelm* in 1931, when he was a first-year student at the University of Munich, and his brothers followed suit.

Beyond the nationalistic ideals of the *Stalhelm*, however, young Josef seemed to have had no other clear political allegiances, and certainly none to Nazism. His closest childhood friend, Hermann Lieb, insists that Mengele was much more interested in dances, girls, and swimming parties in the Danube than in the political movements of the day.

Deeply ambitious, Mengele desired to make a name for himself—not simply to inherit his father's. Although an indifferent student, Mengele knew he wanted to escape a life in the farm-equipment factory. The family business held no long-term interest for him. He would strike out on his own, away from Günzburg and the shadow of his competent father and his domineering mother. Though his grades were uniformly poor, his goal was lofty—he would become a doctor and pursue studies in anthropology and genetics. With chilling prescience, he told a friend at the Gymnasium, "One day, my name will be in the encyclopedia."

Since bad grades were not a deterrent to entering German universities, in the autumn of 1930 Mengele left Günzburg to begin his studies at the University of Munich. The townspeople of Günzburg were genuinely sorry to see him go, as were the workers at his father's factory. Of the three Mengele sons, Beppo was their favorite. He had been, as they liked to put it, "the best of the Mengeles," the friendliest and the most humane.

The university was worlds apart from what Mengele had known in provincial Günzburg and his stuffy boyhood Gymnasium. Munich was a bustling, cosmopolitan city, filled with cultural outlets. Mengele, a longtime lover of opera and classical music, had his pick of concerts to attend. There were also elegant shops, fine restaurants, art galleries, and the Pinakothek, one of the best museums in the world.

Mengele's freshman year, 1930, was also a time when Hitler was beginning to find fertile ground for his most radical ideas. With millions unemployed due to the economic depression, vast crowds were gathering at his rallies. Mengele could not have been oblivious to the speeches that were luring so many of his countrymen, filled as they were with promises of a new Germany, cleansed of all elements that had led to its decline, and so powerful it would conquer the world.

The centerpiece of Hitler's speeches was a call for "racial purity," an idea that was to become the driving force of Mengele's existence. The future dictator beguiled his audiences with his dream—a country populated by blond, blue-eyed supermen and superwomen, a vision that would be achieved through a strict program of "racial hygiene." Hitler called for the elimination of all "inferior" races, especially the Jews.

As Hitler gathered strength, the traditional academic disciplines began to mirror his obsession with race. Even as Hitler and his supporters pressed the need to stop the "contamination" of Germany with Jewish blood, Munich university professors expounded similar notions in the classrooms with great shows of profundity. Subjects such as genetics, biology, history, and anthropology were all being taught with a markedly anti-Semitic slant.

Mengele's chosen fields of anthropology and genetics were especially influenced by the racist theories of Nazi dogma. While the Brownshirts and storm troopers tried to realize Hitler's dream of a new Germany through violence and terror in the streets, genetic scientists were hard at work inside their laboratories, buttressing his theories on the racial inferiority of Jews.

At the University of Munich, Mengele metamorphosed from a carefree bon vivant into a serious student of ideas. It was in Munich that Mengele first exhibited the obsessive attention to work that would mark his later years. At twenty, Mengele displayed more ambition than ever before, as the influence of his hard-driving father—and self-denying mother—increasingly revealed itself. At the university, Mengele was discovering an outlet for his burning need to succeed. His university records suggest that he worked very hard and carried a heavy course load compared to his peers. Friends from the period also recall how diligent Mengele was. Whenever they dropped in to see him at his apartment, he would greet them with a book in hand. Indeed, the ideas Mengele was so anxiously absorbing in his studies were precisely the ones that would propel him down the road to Auschwitz. His apprenticeship as a mass murderer formally began not on the selection lines of the concentration camp but in the classrooms of the University of Munich.

As his autobiographical writings indicate, Mengele was beguiled by the notion of creating a superior race; it was the catalyst for his decision to become a eugenic scientist, in addition to obtaining a medical degree. To improve his credentials as a racial doctor, he sought a Ph.D. in anthropology, and wrote a lengthy dissertation. The possibility of molding a perfect breed of Aryan gods struck a responsive chord in the boy from Bavaria with the aristocratic pretensions. The vanity and sense of superiority had always been there, even in the genial Beppo of Günzburg. What the villagers and the workers at his father's factory had seen as friendliness and bonhomie were cut from the same cloth as the deceptively charming demeanor Mengele would maintain throughout his life, right up to his apotheosis as the "Angel" of Death.

Early on in his studies, Mengele had been introduced to the work of the social Darwinists, who nearly a half-century earlier, in Victorian England, had argued that "biology is destiny." The social Darwinists had believed that nearly all personal and social problems were inherited. Alcoholism, insanity, even poverty and left-handedness, were the result of bad genes. The social Darwinists espoused a program of active intervention to ensure that only the "best" people survived. They wanted to encourage the genetically fit to have more children, while those of questionable stock would remain childless.

The theories of the social Darwinists evolved and gained credence, so that by the 1920s several countries, including Germany and the United States, were fostering a eugenics movement. In Washington, for instance, as Congress was considering new immigration laws, several members advocated the need to limit entry of Eastern European refugees on the grounds they would "contaminate American blood." While arguing for passage of the 1924 Immigration Act, some lawmakers maintained that Latin Americans, Slavs, and even the Irish were "mentally and morally inferior to Nordics."

There was an idealistic component to the eugenics movement, which flourished in Germany long before Hitler. German scientists were not that different from their American and English counterparts in promoting their programs of "racial hygiene." Eugenicists everywhere saw themselves as visionaries, idealists working for a better world through the human gene. They were to create a Utopia, a world free of poverty, illness, and all other physical and mental handicaps. The Germans, however, carried racial science further than did theorists of other nations, and their eugenics movement was above all else distinguished by its extreme anti-Semitic component.

The messianic quality of social Darwinism seems to have appealed to the young Mengele. His writings suggest that he was especially struck by their use of the phrase "the fate of mankind." From his youthful encounter with their distorted ideals, to his old age, a weary and broken exile, Mengele would continue to feel a personal allegiance to the social Darwinists. At the university, the question of the "biological quality of mankind" may have been esoteric to most of Mengele's classmates. But for him, it was apparently a clarion call. His account of that period suggests he was deeply upset by the fact that the lower classes were having many children, while those of impeccable genetic stock were too busy even to marry.

In 1935, the year Mengele prepared to graduate, the Nazis laid down the cornerstone of their racial program, the Nuremberg Laws. The laws introduced complex, ostensibly scientific criteria to determine who was a true German and who was not. According to those laws, Jews were no longer German citizens—citizenship in the Reich being an honor conferred only on the "racially pure." Marriages and sexual relations between Jews and "people of German blood" were now forbidden and carried severe penalties, including lengthy jail terms. Jews were expelled from all professions, including academia. They could not even employ German servants.

Many of Germany's top scientists and research centers joined forces with the Nazis and actively helped to implement the Nuremberg Laws. But none did so with more fervor than the University of Frankfurt's Professor Otmar von Verschuer, possibly the most acclaimed racial scientist of his day. Since the early 1930s, Verschuer had been receiving generous financial support from the Nazis to build his lavish new Institute for Heredity, Biology, and Racial Purity, which specialized in racial studies. Under his able leadership, the Frankfurt Institute, with its impressive staff that included a chief doctor, two medical assistants, and several technicians, did as much work to advance Nazi political aims as it did genuine scientific research. Doctors and scientists from the institute were used by the Nazis as "expert witnesses" in court to prosecute anyone who broke the Nuremberg Laws in their dealings with Jews.

At twenty-five, Mengele graduated from Munich with the highest honors, *summa cum laude*. Now both physician and *Doktorfater*, he contemplated his next step up the academic ladder. That would have to be Verschuer's institute at the University of Frankfurt. He promptly applied for and was given a position working as an assistant to Verschuer, the rising star of Germany's eugenics movement.

Frankfurt in the 1930s was a showplace for the administration of the new Nazi regime. The city's ministry of health set up a special unit just to investigate its residents to determine whether any of them were secretly tainted with Jewish blood or carried any other "genetic disease." Nearly half the citizens of Frankfurt were on file, their dossiers bulging with detailed information on their personal backgrounds.

One of Mengele's duties at the institute was to help in this evaluation of the "racial fitness" of Frankfurt's citizens—really to determine whether their Aryan ancestry was pure. Under the Nuremberg Laws, even individuals with distant Jewish relatives were considered Jewish. Mengele joined his institute colleagues in helping to assist the judicial investigations into violations of the law by determining whether a person charged was indeed a Jew. Even in the years before deportations to the camps began, a determination by Mengele that individuals were racially Jewish could jeopardize their life in Germany, by stripping them of their livelihood as well as depriving them of their standing and security. Within a few years, Mengele's judgments as to who was and was not a Jew, combined with the thousands of dossiers compiled by other Frankfurt bureaucrats, would prove most useful to the Nazis when they were seeking out all those with Jewish ancestry for deportation to the death camps.

JUDITH YAGUDAH:

It was like a nightmare when we arrived. We could see the chimneys, and the flames pouring out of these chimneys. My father turned to my mother and said, "You see, Rosie, the Germans are taking us to burn us."

I was ten years old when my family was deported to Auschwitz.

Just one week before, my twin sister, Ruthie, and I had celebrated our tenth birthday inside the Cluj Ghetto.

I grew up in a middle-class family. I was born in Brassov, a small town in Transylvania. My father worked as a clerk. My mother, Rosie, was a housewife. Ruthie and I were their only children; we were identical twins.

Mother's entire life was centered around us. She adored Ruthie and me—especially Ruthie, who was the livelier twin.

In May 1944, all the Jews in our town were taken to Cluj and forced to live inside a ghetto. We went straight from this ghetto to Auschwitz.

All the Jews of the Cluj Ghetto readily boarded the transports. We had no idea we were going to an extermination camp. No one in our town knew.

Our local Jewish Council—the Judenrat *—had made a secret pact with the Germans. The council had arranged for the Nazis to send us postcards from other Jews who had been deported to the death camps. In these postcards, people we had known wrote that they were fine. They said they had been taken to some labor camps in Hungary, and were working.*

And so my father had no idea what Auschwitz was when we arrived. But when we got out of the cattle car, and he saw the dogs, and the Nazis in uniform, and the flames billowing out from the crematorium, he guessed.

We were immediately separated from our father. Ruthie and I went with our mother to one line just for women and children.

There was Mengele, standing at the head of the line.

He was telling people where to go, in what direction—to the right or to the left, to work or to the gas chambers.

When it was our turn, Mengele immediately asked us if we were twins. Ruthie and I looked identical. We had similar hairdos. We were wearing the same outfits.

Mengele ordered us to go in a certain direction—and our mother, too.

With his medical training and Ph.D. in anthropology, Mengele was well-prepared to excel at the Frankfurt Institute. His job as an assistant to Verschuer marked his formal initiation into the world of Nazi racial medicine. He quickly rose to become Verschuer's most trusted young disciple, and they often collaborated on important projects. Verschuer clearly saw potential in the malleable young man who was not only hardworking and determined, but whose ideas were also politically congenial to Verschuer's own. Under his aegis, Mengele learned that it was acceptable—even desirable—to experiment on human beings if it advanced a scientific cause. The professor provided the critical link in Mengele's transformation from an ambitious young scientist into a camp doctor who sent Jews to their deaths and performed grotesque experiments on children.

Indeed, the obsession with twins that Mengele would later exhibit at Auschwitz was also a direct result of his association with Verschuer. In the 1920s, when Verschuer had been head of the genetics department at Berlin's Kaiser-Wilhelm Institute, the leading research center in Germany, his work had focused almost exclusively on twins. Mengele's mentor was convinced that twins held the key to unlocking the mysteries of genetics. He wrote in one textbook that experimenting on twins would enable scientists to make "a reliable determination of what is hereditary in man." Verschuer, however, is not known to have conducted any so-called "in vivo" experiments on living twins. He restrained himself either because of long-standing rules forbidding the use of human guinea pigs that traditionally bound the scientific community, or because of his own devoutly Catholic upbringing. Until the war removed such protocols, his work, instead, like that of so many other researchers on twins, was based on observation and patient comparisons. Verschuer did, however, instill in his young assistant his confidence of how useful twins' studies, including "in vivo" research, could be.

EVA MOZES:

During the journey to Auschwitz, our father had gathered the family near him in a corner of the cattle car. "Promise me that if any of you survive this terrible war, you will go to Palestine," he told us.

We arrived at Auschwitz in the early spring of 1944. There were six of us: my mother and my father, my older sisters, Edith and Alice, and me and my twin, Miriam.

My father, Alexander Mozes, was a very religious man. He opened his prayer book and began to pray—right there, in the crowded cattle car. But since Jews have to pray facing east, he had to pause for a moment to figure out in which direction he should turn.

He opened his prayer book and began calmly to read amid the cries of hungry children and their terrified parents. A few others in our car joined him in the recitation of the Shema, *the ritual Hebrew morning prayer.*

When the doors to our cattle car opened, I heard SS soldiers yelling, "Schnell! Schnell!" *("Faster! Faster!"), and ordering everybody out.*

My mother grabbed Miriam and me by the hand. She was always trying to protect us because we were the youngest.

Everything was moving very fast, and as I looked around, I noticed my father and my two older sisters were gone.

As I clutched my mother's hand, an SS man hurried by shouting, "Twins! Twins!" He stopped to look at us. Miriam and I looked very much alike. We were wearing similar clothes.

"Are they twins?" he asked my mother. "Is that good?" she replied. He nodded yes.

"They are twins," she said.

❧

LEA LORINCZI:

My twin brother, Menashe, and I were vacationing with our grandparents in Transylvania when the Germans came in March 1944. My grandparents tried to send us home to our mother in Cluj. But we were not permitted to travel.

The police rounded us up and placed us in a ghetto. Menashe and I celebrated our tenth birthday inside this ghetto. As a birthday present, our grandmother gave each of us one slice of bread.

From the ghetto, we were all placed in cattle cars and taken to Auschwitz. When we got off the trains, we could hear the Germans yelling, "Twins, twins!"

My grandmother naively believed our mother was there, and had instructed the guards to be "on the lookout" for us. She thought that was why they were calling for twins.

And so Grandmother pushed us out of the line going to the gas chamber and said, "You are going to your mother." She thrust into our hands a toothbrush and toothpaste.

As Mengele was absorbing Verschuer's values, he was also working extremely hard to please his mentor. As a result of his diligence, Mengele became not only a special pet of Verschuer at work but also a favored guest at the director's residence, where he often stayed for dinner.

There were personal factors that strengthened the bond between the young would-be scientist and the dean of Nazi racial hygiene. Mengele was far from home, and deeply in need of a parental figure to give him support. While in Günzburg, Mengele had suffered from his father's self-absorption, the fact that he was more involved in his factory than with his sons. Mengele was naturally drawn to Otmar von Verschuer, who was a personable, fatherly man with several children of his own, as well as a brilliant scientist. In establishing a close tie with Verschuer, Mengele found both the paternal attention he longed for and reaffirmation of his own talents.

In many ways, the impressionable Mengele was an ideal protégé for the opportunistic Verschuer. The professor could and did channel Mengele's zeal to distinguish himself to advance the institute and, in turn, the cause of the Nazis. Mengele's mixture of intense vanity and insecurity made him ripe for manipulation. Young Mengele would be groomed to be a "biological soldier" who would obey Verschuer's orders in the laboratory as completely as a soldier in the battlefield.

While Mengele toiled away in the laboratory, his politically savvy mentor was spending time currying favor with Germany's new rulers. Verschuer, who was in constant communication with the Nazi hierarchy, paid frequent tribute to Hitler in his various publications. His articles routinely praised the Führer, damned the Jews, and called for ever more drastic measures to eliminate them from German society. Verschuer's anti-Semitic rhetoric grew more vivid with the increasing power of the Nazi regime.

Ultimately, Verschuer helped give the Final Solution—the plan to kill all the Jews of Europe—intellectual respectability. Although he never held high office in the Reich, his published opinions carried considerable weight. The Nazis relied on him to offer scientific rationales for their more brutal actions. Not coincidentally, the more he endorsed the Nazi line, the greater became his prestige and influence. As a perceptive American investigator at the Nuremberg trials later observed, Verschuer "sacrificed his pure scientific knowledge in order to secure for himself the applause and the favor of the Nazi tyrants."

By 1938, Mengele was beginning to enjoy considerable professional success and recognition. He had just completed and published his second doctoral dissertation, which was along similar lines of his earlier Munich treatise, a study of the general area of the human jaw. This new thesis clearly reflected Verschuer's influence. While the earlier paper had been primarily a factual study, Mengele now began to formulate theories on the "racial origins" of hereditary traits such as the cleft palate and the harelip.

Mengele's personal life, once limited to group outings and casual flirtations, was also flourishing. That same year he became engaged to Irene Schoenbein, a lovely young German woman he had met while on holiday. But even as they set a wedding date, Mengele was summoned for a three-month stint in the Wehrmacht, the German Army. Preparations for war were under way, and young men around the country were being called up for training. Mengele was assigned to a mountain regiment in the Tyrol—seemingly an ideal assignment for the young man who enjoyed skiing and hiking.

But Mengele's experience in the Wehrmacht proved to have a fateful impact on his life. According to Dr. Kurt Lambertz, one of his best friends from the period, Mengele developed an intense dislike of the unit's commanding officer. The personality clash ended in a brawl between the two. Although Mengele completed his training period, his career in the Wehrmacht was finished. He decided instead to join the SS—the Nazis' most elite and ruthless corps of soldiers. As a young officer/doctor, Mengele was assured of a distinguished future, serving with the brightest, the most dashing—and the most fanatic—Nazis.

Mengele and Irene were married in July 1939. Weeks later, war broke out and Mengele found himself drafted. His first assignment was an administrative post at the SS Race and Resettlement Office, reviewing applications for German citizenship. In the countries surrounding Germany, such as the Soviet Union and other Slavic lands, there were hundreds of thousands of people who claimed German ancestry. In Hitler's view of the world, the German "nation" encompassed all Germans—wherever they might be living. It was the duty of the Reich to bring them back into the fold. Mengele's office was charged with sorting through the applications to determine the "real" Germans: those who met the racial and genealogical criteria.

It was almost two years before Mengele got his first taste of combat. In June 1941, he was sent to the Ukraine as part of the Waffen SS. He proved to be an excellent soldier, and received the Iron Cross, Second Class, for his heroic service on the battlefield. The following year, he joined the SS Viking Division as a field physician. Here, Mengele finally practiced medicine, but under the worst conditions. Epidemics were common during the hot, sticky summers. Winters were so cold as to be unbearable. Thousands of men died every day in fierce battles, and there was neither enough equipment nor medication to keep the wounded alive.

It was on the Russian front that Mengele honed the art of selection. Due to the shortage of time and supplies, he was forced to make snap decisions as to who among the wounded would be treated, and who would be left to die. The task of choosing among the German soldiers was gruesome to Mengele, and he hated it, he later told friends and colleagues. But he resolved to be dedicated and brave. He ended up being awarded the Iron Cross, First Class, for pulling two wounded soldiers from a burning tank under enemy fire. As one commendation he received stated, Mengele had "conducted himself brilliantly in the face of the enemy."

TWINS' FATHER (ZVI SPIEGEL):

"Were you ever a soldier?" Mengele asked me after he had pulled me out of the selection line.

I had been standing with a group of twins. Because of my years as an officer, I tended to stand very straight. Mengele noticed that immediately.

I told Mengele about my background in the Czech military. Because of this, Mengele appointed me to be in charge of the twin boys. My title would be "Twins' Father." I was to supervise about eighty boy twins.

But he warned me that if anything went wrong, I would be killed on the spot.

At the end of 1942, Mengele was wounded—it is unclear how—and declared unfit for combat. He was sent back to Germany and reunited with his wife and his old mentor, Verschuer. The SS reassigned him to their Race and Resettlement Office, working in the Berlin Division overseeing the concentration camps. It was another desk job, far from the scientific work he loved.

Sometime that year, Verschuer had left Frankfurt to assume the lofty position of director of the Kaiser-Wilhelm Institute in Berlin, possibly the most prestigious scientific job in Germany. As Mengele grew disillusioned with his bureaucratic job, Verschuer and he began exploring other options that would get him back where he wanted to be: inside a laboratory.

Both Mengele and Verschuer were aware of the exciting research being undertaken in some of the concentration camps. Since 1939, medical-research projects of various kinds had been under way, including experiments performed on human subjects. All known standards of medical ethics had been swept aside by the Nazis. To find a cure for typhoid fever, the bane of the German Army, Nazi doctors infected prisoners at the Buchenwald camp with the virus, then tried to "treat" them by injecting them with various serums. At Dachau and other camps, Jews and other inmates were exposed to tuberculosis, cholera, diphtheria, smallpox, influenza, and yellow fever as the Nazis tried to learn how to control these deadly maladies. Nineteen Forty-two was also the year when German doctors began their gruesome attempts to discover the most efficient method of mass sterilization.

By far the most intriguing research possibilities were offered at the Auschwitz concentration camp in Poland. It was at Auschwitz that a certain Dr. Horst Schumann had begun exposing men, women, and children to massive doses of radiation—a promising way of achieving the desired sterilization. The largest of all the camps, Auschwitz had more than ten thousand inmates arriving each day—an unimagined number of potential human subjects. They could provide a scientist with a broad cross-section of racial groups. Twins and other interesting genetic specimens were likely to pass through as well.

MAGDA SPIEGEL:

I was twenty-nine years old, married, with a son of my own, when my family was deported to Auschwitz from our village in Czechoslovakia.

We had all made the trip in the same cattle car, but we were separated the minute we arrived. My mother, my son, and I were told to go to the left side, toward the crematoriums.

SS guards were yelling, "Twins, twins, we want twins." I saw a very good-looking man coming toward me. It was Mengele.

He was with two guards, and my twin brother, Zvi. My brother had told Mengele he was a twin, and that he had a sister.

❧

TWINS' FATHER:

To this day, I am not sure why I admitted I was a twin. My previous experience in labor camps had taught me never to volunteer for anything.

Magda and I were born in 1915 in Budapest. When we were little, our family had moved to Munkács, a town in Czechoslovakia renowned throughout Europe for its flourishing Jewish community.

When I was twenty-one, I was drafted into the Czech military, [where] I became an officer. But because I was Jewish, I was ultimately sent to a series of labor camps. When I came back to my hometown, it was only to be placed in the Jewish ghetto.

From there, my entire family—my parents, my twin sister, and her son—were put in a cattle car bound for Auschwitz.

Once I had informed Mengele I had a twin sister, he went looking for her—and plucked her out of the lines marching to the gas chambers.

❧

MAGDA SPIEGEL:

Mengele pointed to Zvi and asked me, "Are you the twin of this man?" I said yes.

Then, Mengele noticed my child. "Who is this little boy?" he asked. "He is my son," I answered.

"Please leave the boy with your mother," Mengele told me very nicely.

With the aid of Verschuer, Mengele obtained a position as an SS doctor at Auschwitz. Verschuer even helped Mengele win grants to undertake two research projects at the camp. He was to begin in April 1943.

What a splendid laboratory Auschwitz promised to be! Unique in the world of science, it offered unlimited possibilities for work . . . for medicine . . . for experiments. At last, the chance to do the kind of research Mengele had dreamed of. At Auschwitz, there would be nothing to stand in his way.

HEDVAH AND LEAH STERN:

Mother was determined to hold on to us. She hid us under her skirt.

But at the last minute, she told us, "Go to Dr. Mengele. He is asking for twins. Go and we will meet by the gate.

"Wait for me, children, wait for me," she cried. "We will meet again by the gate."

2

AUSCHWITZ MOVIE

MAGDA SPIEGEL:

A few hours after arriving at Auschwitz, I asked some people, "Where is my little boy?" My son was only seven years old. I was very worried about him.

"You see these chimneys?" they replied, pointing toward the crematoriums. "Your child is there. Your parents are there. Your entire family is there. And one day, you will also be there."

This was told to me the same day I had come—the same day.

Dr. Mengele was the only person who was always standing there when the trains came. He was constantly making selections.

The sky was red—red—the whole sky was red!

It was the last year of the transports, and the Germans were putting masses—masses and masses—of people into the crematorium.

It was like watching a movie.

Even early in the morning, the sky over Auschwitz looked opaque and foreboding, as if it were covered by a vast blood-soaked sheet. An oppressive smell permeated the air—soot and burning flesh, fumes from the crematoriums, and smoke from the arriving trains.

After the trains had pulled in and the cattle-car doors were opened, exhausted cargoes of Jews tumbled out. As SS men shouted, "Faster! Faster!," hordes of people were pushed this way and that by the uniformed guards. Women cried as their husbands were taken away from them. Old men clutched their wives in a final embrace. Small children huddled closer to their parents, sad and subdued. And the Nazis stomped around, cracking their whips on anyone who stood in their way, and even on those who were merely standing.

VERA BLAU:

When I arrived at Auschwitz in April 1944, my first impression was that it was very crowded.

My twin sister, Rachel, and I were eleven years old. We had come with our mother and little brother, and both of us started crying when we were separated from them. Then a woman from Czechoslovakia came over to us. She had been in Auschwitz a long time.

"Do not cry, children, do not cry," she said to us. "You see, they are burning your parents."

I did not believe her. I did not want to believe her.

What is universally known today as Auschwitz is in fact something of a misnomer. Auschwitz was the slave-labor camp in which murder was an everyday phenomenon, but, in fact, the Polish place name became the umbrella word for several camps. Although the slaves largely labored at Auschwitz, it was at Birkenau, a couple of miles away, that many of them were executed. And although the world lexicon came to equate Auschwitz with the gas chambers, it was Birkenau that was the actual extermination center. It was Birkenau where the ovens never stopped flaming and where SS physicians regularly dispatched inmates to the crematorium; and it was Birkenau where Dr. Mengele worked in his laboratory, and where his beloved twins were barracked, and where so many of them inevitably perished.

Just one year after arriving at the death camp, Mengele was thoroughly absorbed in his research, the first step of which was selecting his subjects. Every morning, at the crack of dawn, he could be seen in the area where the transports disembarked, scanning the new arrivals.

Standing there in his perfectly tailored SS uniform, white gloves, and officer's cap, Mengele looked impeccable—a host greeting guests arriving at his home. He sometimes stood for hours without flinching, a hint of a smile on his face, his elegantly gloved hand beckoning the prisoners to the right or to the left. Often, he whistled softly as he worked, the *Blue Danube* waltz, or an aria from his favorite Puccini opera.

Mengele even engaged some of the new arrivals in friendly conversation, asking them how the journey had been, and how they were feeling. If they complained of being sick, he listened with a sympathetic ear—and then sent them straightaway to die in the gas chambers. He actually seemed interested in hearing all the gruesome details: how uncomfortable the trip had been, how cramped and stifling the cattle cars were, how many Jews had died along the way.

Occasionally, Mengele pulled aside inmates and asked them to write "postcards" to their relatives back home. He seemed to take a special pleasure in dictating these notes, describing how lovely Auschwitz was, and urging everyone to visit. But once the postcards were prepared, their authors were summarily dispatched to the gas chambers.

Only when an interesting "specimen" came along did Mengele really spring to life. He urgently motioned to a nearby guard to yank the new arrival out of the line. SS guards were ordered to watch for any unusual or striking genetic material—the dwarfs, the giants, the hunchbacks—and to bring them immediately to Mengele. But most important of all to him were the twins.

ZVI THE SAILOR:

My twin brother and I were marching toward the gas chambers when we heard people yelling, "Twins! twins!" We were yanked out of the lines and brought over to Dr. Mengele.

I was not quite thirteen years old when my family was deported to Auschwitz—I hadn't been bar mitzvahed yet. I came from a small village in Hungary where my father's family had lived for generations. My mother came from Galicia, in Poland. There were eight children in our family.

When we stepped off the cattle car, there was Dr. Mengele.

He was making the selections, deciding who would go to work and who would go to the gas chambers. He used his finger. He motioned everyone in my family in the direction of the crematorium.

As we marched to the crematorium, our mother told us, "You must not cry."

To this day, I do not know who told the Germans we were twins and had us removed from the line.

❧

MENASHE LORINCZI:

Nobody knew whether it was good or bad to be a twin. Although the SS guards were going around asking for twins, families were afraid to volunteer their children.

Many twins died because their parents didn't want to be separated from them. Mothers walked with their twins straight into the gas chambers.

❧

EVA MOZES:

Once the SS guard knew we were twins, Miriam and I were taken away from our mother, without any warning or explanation.

Our screams fell on deaf ears. I remember looking back and seeing my mother's arms stretched out in despair as we were led away by a soldier.

That was the last time I ever saw her.

The twins who passed through the gates of Auschwitz were of all ages, but often they were very young children who fought and cried at being separated from their loved ones. If Mengele was on the scene, he tried to soothe the terrified parents. He would smile as he comforted an anguished mother, insisting her twins would be in good hands. And if the twins were just infants, Mengele might sometimes pull their mother out of the line as well, permitting her to accompany and look after them. Most often, however, the children were taken away alone.

Once separated from their parents, the twins were marched through the camp, where they witnessed scenes of unparalleled horror. Piles of corpses were everywhere. Lying next to them, and virtually indistinguishable, were men and women thin as skeletons. These were the "Mussulmans"—the half-dead, with no strength or will to live, who were simply awaiting being carted to the gas chamber. A foul odor permeated the camp, which, combined with the heat, made it difficult to breathe. It was an absolute assault on the senses. Children clung to their twin, their last remaining links with the families they had lost.

The twins' initiation into Auschwitz formally began when they, like all new inmates, were showered and branded. They cried out in pain as numbers were etched into their flesh with searing metal rods. But unlike the other prisoners, who were given camp uniforms and whose heads were shaved, the twins were allowed to keep both their clothes and their long hair. These differences made them immediately recognizable as "Mengele's children."

Despite these small privileges, the twins sank into despair within hours of arrival as they began to understand what had happened to their families. Once in their own compound, where at any given time there could be scores of twins, boys and girls separately lodged, they were briefed by the other children about the realities of life and death at Auschwitz-Birkenau. Those newcomers who had not understood what they had seen were told about the gas chambers and crematoriums, and the probable fate of the family members they had left behind. In the case of male twins, whose wooden barracks stood only yards away from the crematorium, virtually facing it, Twins' Father took it upon himself to break the news gently, and at times delayed it for days or weeks. The little girls, who had no such parental figure to ease the transition, were less fortunate. Even though many of the children chose not to accept, or were too young to fully comprehend, what they were told about their own parents, it was a devastating moment.

MOSHE OFFER:

I felt so tired, that first day at Auschwitz. There was a terrible smell —it was impossible to escape the smell.

I was very worried about my mother, my father, and my four brothers. I talked with [my twin brother] Tibi about them.

But he was sure our mother was going to be safe.

❧

HEDVAH AND LEAH STERN:

We kept crying and looking for our mother. She had promised she would meet us at the gate.

We would search among the women for her dress. When we were separated, she'd been wearing a striking black dress with pink strawberries.

We couldn't eat. We were constantly crying and looking toward the gate for our mother.

Finally, the head of our barracks said, "Come here" and pointed to the crematorium.

"I can now tell you that your mother and the rest of your family went to the gas chambers."

❧

EVA MOZES:

In the early evening, we were finally taken to our barracks. There, we met other twins, some of whom had been at Auschwitz a long time.

There were only girls in the barracks, I can't remember exactly how many. Maybe hundreds of little girls. The barracks themselves were filthy. They had these red brick ovens [for heating] running across them, and wooden bunk beds, without pillows. [We] slept two, three, four girls to a bunk bed.

That first night, we went to the latrines. They were just holes in the ground, with waste in them. There was no running water. Everything stank.

I remember seeing three dead children on the ground. Later, we would always be finding dead children on the floor of the latrines.

From our barracks, we could see huge smoking chimneys towering high above the camp. There were glowing flames rising from above them.

"What are they burning so late in the evening?" I asked the other children.

"The Germans are burning people," they answered.

But the new twins also learned that, as protégés of the powerful Dr. Mengele, their own lives in this kingdom of death were guaranteed. Mengele made sure that his twins would be generally well-treated, at least by Auschwitz standards. They were spared the beatings and punishments inflicted on other inmates. Because they "belonged" to Mengele, no one, not even the most brutal camp guards, would dare lay a hand on them. In addition to keeping their clothes and hair, some of the twins, especially the boys, recall receiving somewhat better food rations than the other prisoners. Although all the twins say they were ravenously hungry throughout their stay, several remember having access to potatoes and slices of bread, which enabled them to survive. If caught stealing food—as many did, on a regular basis—they were not severely punished because of their protected status. Most important, the twins were not subjected to the terrifying random selections that adult prisoners faced. As long as they stayed healthy and useful to Dr. Mengele, they would be kept alive.

The work habits Mengele had developed over the years in Munich and Frankfurt stood him in good stead at Auschwitz. Since his arrival in May 1943, Mengele had distinguished himself in the eyes of the Nazi hierarchy. A superior's evaluation praised him for being "an excellent officer," who had shown "maturity and strength." The report stressed how Mengele had not displayed "any weakness in character or inhibition," in the resolute way he selected people to die. And one doctor who served with Mengele at Auschwitz, Dr. Munch, remembers him as much more diligent than other SS physicians, many of whom had been dragooned into service at the death camp. This perception of Mengele as more hardworking than his Nazi colleagues is echoed by numerous adult survivors who had the chance to observe him at close range, and who would go on to write about it in their memoirs and testimonies. Mengele's experimental barracks were a showcase of the concentration camp, talked about and admired by the Nazi hierarchy.

In many ways, the twins' compound at Auschwitz was a realization of Mengele's—and Verschuer's—greatest scientific dream. Verschuer, from his position as director of the Kaiser-Wilhelm Institute in Berlin, was closely involved in his protégé's research, and the two men corresponded regularly. Mengele periodically dispatched to his mentor not only reports about his research, but also laboratory samples from his experiments.

TWINS' FATHER:

The moment a pair of twins arrived in the barrack, they were asked to complete a detailed questionnaire from the Kaiser-Wilhelm Institute in Berlin. One of my duties as Twins' Father was to help them fill it out, especially the little ones, who couldn't read or write.

These forms contained dozens of detailed questions related to a child's background, health, and physical characteristics. They asked for the age, weight, and height of the children, their eye color and the color of their hair. They were promptly mailed to Berlin when they were completed.

After the form was filled out, I would take the twins to Mengele, who asked them additional questions. Mengele had an office and a beautiful blond secretary. Because many of the children only knew Hungarian—and Mengele spoke only German—I would serve as the translator. His gorgeous secretary would write it all down.

Much of the information Mengele was seeking had to do with demographics—where the family was from, what the parents had done for a living.

His secretary would measure the children, while Mengele examined them. He was especially interested in their hair. I recall he would look closely at the roots, to see how it was growing.

The questions were connected to the experiments Mengele would later make on the twins.

One day, I was filling out forms for a new pair of twins and I noticed the date of birth one child had given me was different from the birth date of his sibling. It was obvious they were not really twins. But I knew that if anyone learned of this, the boys would immediately be put to death.

And so I decided to take a chance, and put down false information. I "made" them twins. I knew if Mengele learned of what I had done, he would kill both me and the children on the spot.

And throughout my stay at the camp, I was always afraid—but also secretly delighted—at what I had done in slipping through these false twins. I felt I had tricked the great Dr. Mengele.

The detailed questionnaires were designed to ensure the validity of Mengele's work. If the experiments were to have any relevance, they had to be precise and carefully controlled. While Nazi racial scientists considered identical twins, whose gene pool was exactly alike, as most desirable for study, fraternal twins were also useful. Whatever the case, the more information gathered about a twin's genetic background, the greater the chances of conducting meaningful experiments.

Mengele's overall aim—and that of Verschuer—was to test various genetic theories in support of Hitler's racial dogmas. Like other Nazi scientists, Mengele hoped to prove that most human characteristics, from the shape of the nose to the color of the eye to obesity and left-handedness, were inherited. In addition, it is believed Mengele was searching for ways to induce multiple births, so as to repopulate the depleted German Army. The ultimate goal was to produce an ideal race of Aryan men and women endowed with only the finest genetic traits, who would rapidly multiply and rule the world.

For the sake of his experiments, Mengele tried to create an atmosphere that was, in sharp contrast to everything else taking place in Auschwitz, as close to normal as possible. He installed a small furnished office at one end of the twins' compound to help him monitor his child guinea pigs. He decreed that guards were to be held accountable if any of the twins fell ill or died, hoping this would motivate them to look after the children. If a twin died during the night, Mengele would storm through the compound in the morning, screaming at the guards, demanding an explanation. He also implemented a strict routine to regulate the twins' lives. Every morning at six o'clock, they were to be up for roll call in front of their barracks. Then came breakfast, a mug of tepid, muddied water the Germans called coffee, and perhaps a slice of moldy bread. Then Mengele would appear at the compound shortly thereafter for an inspection tour.

EVA MOZES:

My first meeting with Dr. Mengele was the morning after I had arrived.

The twins had to stand on roll call, no matter how young they were, no matter how cold it was. The procedure could last anywhere from fifteen minutes to over an hour.

The Germans had to account for everyone. Once that was finished, Mengele came. He was very much like a general reviewing his troops —except we were his guinea pigs.

I remember Mengele came almost every day, and he always wore his SS uniform and tall black boots. They were very shiny boots.

I was terrified of him.

As an SS doctor, Mengele had many concerns and responsibilities in addition to the twins. He performed selections among the new arrivals on the Birkenau ramp and also inside the women's compound, daily dispatching hundreds to be killed. He oversaw a kind of sham hospital for the sick, where, alas, little care was provided. But he seemed to spend most of his time with the children. Typically, Mengele gave daily orders to have several of the twins "prepared" for experiments. He would ask Twins' Father or other adult supervisors to get the children ready by taking them to be bathed and cleaned. Special trucks emblazoned with fake Red Cross insignia arrived to pick up the youngsters and deliver them to Mengele's laboratories. Depending on the type of tests, the twins were driven to any one of several locations either within Birkenau or at Auschwitz proper. The children learned what to expect, depending on their destination. In one laboratory, they knew it was a matter only of routine X rays and blood tests. Other clinics were reserved for more complicated—and painful—experiments. One Mengele lab the twins never saw was his pathology unit, located conveniently on the site of a crematorium. There, an assistant to Mengele toiled quietly, performing autopsies on the bodies of the twins who had died, or been killed, in the course of experiments.

The blood tests were the most basic component of Mengele's program. Virtually all the twins were subjected to daily withdrawals of blood. These tests may have been connected to the grant he received from the prestigious German Research Association to study "specific proteins," a project funded at Verschuer's express request. Blood, often in large quantities, was drawn from twins' fingers and arms, and sometimes both their arms simultaneously. The youngest children, whose arms and hands were very small, suffered the most: Blood was drawn from their necks, a painful and frightening procedure. The blood was then analyzed in a special laboratory located near Birkenau.

Although Mengele was invariably present during the experiments, the tests themselves were often administered by his assistants. Typically, these were Jewish inmates who had been doctors and nurses before the war, and had been spared the gas chamber because of their skills. Most had profound misgivings about their work. But they knew that to betray any hesitation in administering an injection or test, even a very gruesome one, would result in their immediate execution. In spite of their anguishing position, a few managed also to alleviate suffering and helped save some lives.

ALEX DEKEL:

I never saw a doctor smiling. They were very depressed, all of them.

I lived in the same place as these doctors. I saw them going through their duties like robots, like machines. They would come back at night to sleep, and wake up early in the morning to report back to the laboratory.

If I ever approached any of them and tried to ask them a question, they would not answer me.

❧

MAGDA SPIEGEL:

There were many Jewish doctors living in our section of the camp—women doctors, men doctors, anthropologists, eye doctors, ear doctors.

Mengele came every day to speak with them and give them orders on what to do with each twin.

There was a beautiful female doctor named Anna. She was originally from Czechoslovakia, and she had been a very famous physician before the war. She was one of the few Jewish doctors who was able to get close to Mengele.

He liked beautiful women, and Dr. Anna was lovely. Dr. Anna was always walking around with Mengele, accompanying him on his rounds.

She was very kind—very compassionate toward the twins. She knew, for example, that I was upset about my son. She tried to comfort me. She would never admit to me he had been gassed.

Finally, she said to me, "Ask Mengele."

I went over to Mengele and asked him as calmly as I could, "Where is my little boy?"

"He is in kindergarten," he replied, smiling. Then, he walked away.

I wanted to die. There was a fence near our barracks—a barbed-wire fence—where inmates would commit suicide. I threw myself on the barbed wire, but some women prisoners ran and pulled me off of it.

Although Mengele's assistants were responsible for administering the tests, Mengele occasionally liked to step in and lend a hand. He looked the twins over carefully, searching for genetic abnormalities or any other unusual conditions. Then, he would demonstrate the "proper" way to insert a syringe or draw blood. He would lecture the Jewish doctors on how to avoid hurting the children. As if he were their old family physician, he brought candy or chocolates along to pacify the youngsters. Mengele knew how to treat children and calm them down so the experiments could proceed. He intuitively understood how much he needed the children's trust for his research to succeed; good results would be obtained only if they were cooperative. But he also knew that the twins, especially the very young ones, were terrified of the procedures, especially the injections.

EVA KUPAS:

Once, I wanted to go see my twin brother. So Dr. Mengele took me by the hand and walked with me over to where he was staying. Mengele held my hand the whole way.

❧

VERA BLAU:

They took blood from us every day. But when Mengele made the blood tests, he was much more gentle than the nurses. He liked little children and was strict about keeping them well. He would get very angry if a twin was sick.

Countless experiments were performed on the twins at any given time. Although the full scope of those tests will never be known, although the records are scanty and the child victims of Mengele's most diabolical work are gone, some details are known about both his practices and intent.

EVA MOZES:

We were always naked during the experiments.

We were marked, painted, measured, observed. Boys and girls were together. It was all so demeaning. There was no place to hide, no place to go.

They compared every part of our body with that of our twin. The tests would last for hours.

And Mengele was always there, supervising.

❧

SOLOMON MALIK:

Mengele would look at each of the twins and see what interested him the most. We were his guinea pigs. We were his laboratory.

Mengele once put a needle in my arm—only the needle, not the syringe. Blood started spurting out. He calmly placed the blood in a test tube.

Then, he gave me a sugar cube.

❧

MOSHE OFFER:

One morning, at roll call, my number and that of Tibi were announced as part of the group that was going for experiments. We were taken with some other children by ambulance to a laboratory. The doctors took many X rays of us.

Then, Dr. Mengele walked in. He was wearing a white gown, but underneath his gown I could see his SS uniform and boots.

He gave me some candy, and then he gave me an injection that was extremely painful.

"Nicht angst," *Mengele told me in German. "Don't be afraid."*

The line between science and quackery was not a very fine one at Auschwitz. Mengele's experiments, although ostensibly performed in the name of scientific truth, followed few scientific principles. Mengele would test one twin and not the other. At times, siblings were injected simultaneously as they stood naked side by side. Despite the preferential treatment, the occasional shows of kindness, the children endured unspeakable pain and humiliation.

The eye studies were especially gruesome. In their desire to create a race of perfect Aryans, the Nazis wanted to produce children with lustrous blond hair and blue eyes. Was it possible to genetically engineer such traits? A major focus of Mengele's studies involved changing the color of the twins' hair and eyes. (His fascination with hair had led him to allow the twins to wear it long. Their hair was continually being analyzed and compared with that of their siblings.)

The eye studies were in part undertaken on behalf of Mengele's Berlin colleague Dr. Karin Magnussen, one of Verschuer's top assistants at the Kaiser-Wilhelm Institute, who was preparing a special study on eyes using data gathered by Mengele at Auschwitz. This data came from painful and barbaric tests that were difficult to administer. At times, Mengele's assistants used eye-drops to insert chemicals into the children's eyes; other times, they used needles.

HEDVAH AND LEAH STERN:

Mengele was trying to change the color of our eyes. One day, we were given eye-drops. Afterwards, we could not see for several days. We thought the Nazis had made us blind.

We were very frightened of the experiments. They took a lot of blood from us. We fainted several times, and the SS guards were very amused.

We were not very developed. The Nazis made us remove our clothes, and then they took photographs of us.

The SS guards would point to us and laugh. We stood naked in front of these young Nazi thugs, shaking from cold and fear, and they laughed.

The strict veil of secrecy imposed over the experiments enabled Mengele to work more effectively. Twins who were subjected to the most grotesque procedures took his secrets to their graves. And those as yet unhurt had only secondhand stories about what was being done. No one at Auschwitz—not the prisoners, not the SS guards, not even other doctors—knew precisely what Mengele was doing with his twins, either while they were alive or after they had died. Even the children who themselves were the subjects of his tests did not know what the objectives of the experiments were. Rumors were rampant, especially when children were taken out in Red Cross trucks, never to be seen again.

There was a calculated effort by Mengele to limit information, presumably to keep his "guinea pigs" from panicking. One can only guess how the twins would have reacted had they known of the existence of Mengele's lugubrious pathology laboratory where bodies of children were daily brought over to his assistants to have still more "tests" performed.

The secrecy enabled Mengele to establish an ostensibly friendly rapport with the twins—at least, the very young ones. He loved to sit and chat with them. He seemed more relaxed, more himself, in the twins' barracks than anywhere else at Auschwitz. Only with those young children could he joke and laugh. With them, he was the Beppo his friends and family in Günzburg had known—ineffably sweet, charming, and affectionate.

Mengele especially seemed to dote on the very youngest twins, the toddlers who could hardly walk and talk, and were completely dependent on him. There was a little boy, about three or four years old, who bore a striking resemblance to Mengele himself. He was a dark child, with large brown eyes, a round face, and a gentle disposition. Mengele came to the barracks often to visit this one child, scooping him up in his arms, kissing him, and showering him with toys and chocolates. When asked his name, the boy would reply, "My name is Mengele." The Auschwitz doctor himself had taught the child to say this, as a father would teach his son.

VERA BLAU:

Mengele loved this child. He had a twin brother, but Mengele played only with him.

I believe Josef Mengele loved children—even though he was a murderer and a killer. Yes! I remember him as a gentle man.

During this period, Mengele himself had become a father. His son, Rolf, was born on March 11, 1944, when Mengele had been at Auschwitz for almost one year. But he could rarely get away from the camp to spend time with Irene and the baby. Mengele may have been lavishing displaced affection for his own son on the little Jewish boy.

The twins were Mengele's pampered children—the most privileged inhabitants of Auschwitz. He arranged more and more perquisites for them. Zvi Spiegel, the Twins' Father, was permitted to hold makeshift classes and recreation periods to entertain the youngsters. An avid sportsman in his own youth, Mengele personally organized soccer matches with teams composed of the twins. It was an eerie sight for other inmates to observe the children running, shouting, tossing a ball.

JUDITH YAGUDAH:

Mengele took us to a concert once. I still remember it—because it was so awful.

It was held just outside our compound. The orchestra was made up entirely of women prisoners. Listening to them play was heartbreaking.

It reminded us so much of normal life . . . the life we'd had before . . . the life that other people still led.

❧

TWINS' FATHER:

I felt very sorry for the young twins. I was always trying to make their lives a little more bearable.

I was twenty-nine years old when I was deported. I felt at least that I had tasted life—whereas most of the twins hadn't even begun to live.

I saw my most important task as maintaining the children's morale. That was difficult, because there were so many periods when I myself despaired of ever surviving the concentration camp.

Quite often, twins would come over to me and start crying. They missed their parents. Or else they were hungry, and their stomachs were hurting.

I would try to comfort them. I calmed them down by telling them they would be meeting up with their parents within a very short time.

The twins wanted more than anything else to go home. They would ask me if I thought the war would ever end. I would reassure them. I promised I would take them home myself the day it was over.

They believed me. They looked up to me, I guess.

I organized classes. I would teach them math, history, geography. We had no books, of course. But I still gave them simple exercises to do. I taught them whatever I could remember from my own school days.

I would also talk to them about the day the war would end and they would all get to go home.

I was very anxious about keeping the children together, and as close to me physically as possible. Children are children wherever they are. The twins wanted to walk around the camp—and, of course, that was very dangerous.

An SS guard who might not know they were Mengele's twins could kill them on the spot. That was why I thought up games and classes. This way, I knew where they were at all times.

I also distrusted Mengele. I felt he could change his mind about the twins at any time and have us all killed. And so my strategy was to have the twins maintain as low a profile as possible—and keep out of Mengele's way.

The youngest children were not aware of the darker side of the man they affectionately called "Uncle Mengele." They did not know about the surgeries of unspeakable horror. It is not that Mengele confined the more gruesome and painful studies to the older children, but that they were too young to comprehend the sinister experiments. They saw only a cheerful, avuncular doctor who rewarded them with candy if they behaved. But the older twins, and the adults, such as the twins' mothers and Jewish doctors who saw Mengele at work, recognized his kindness as a deception—yet another of his perverse experiments, whose aim was to test their mental endurance. The fact that Mengele could behave so nicely even as he inflicted pain, torture, and death made him the most feared man at Auschwitz.

However, in performing his diabolical experiments on twins, Mengele believed he was pursuing high-minded scientific goals. Personal ambition also played its role in propelling Mengele to exceed even Nazi standards of cruelty, so that he stood out from other death-camp doctors. Driven by a need to reach the peak of his profession, he no doubt believed that if he only worked hard enough, performed enough experimental studies, tested a sufficient number of twins, then he would be recognized as the great scientist he thought he was.

But despite his dedication and fanaticism, the mediocre student of Günzburg never possessed any real brilliance. The tests, questionnaires, and many of the experiments themselves appear to have been the brainchild of Verschuer. At his best, Josef Mengele was earnest and efficient, an excellent "assistant," but even at his most fiendish, he was not a man of ideas. In Munich, in Frankfurt, and even now at Auschwitz, he was merely methodical and perverse. If there was ever a "great" racial scientist, it was Otmar von Verschuer.

Mengele's fundamental narcissism, however, led him to try to make himself into a first-rate scientist, as he had made himself elegant and stylish. Mengele never saw his personal limitations and the gap between his real and imagined abilities, of course. He donned genetic theories and questionable racial dogma the way he put on white gloves and a hat. Obsessed with his own appearance, the vain young Beppo of Günzburg was devoting his life to a science whose aims—fostering a race of "beautiful" people—mirrored his own personal obsession.

ALEX DEKEL:

I have never accepted the fact that Mengele himself believed he was doing serious medical work—not from the slipshod way he went about it. He was only exercising his power.

Mengele ran a butcher shop—major surgeries were performed with-

out anesthesia. Once, I witnessed a stomach operation—Mengele was removing pieces from the stomach, but without any anesthesia.

Another time, it was a heart that was removed, again, without anesthesia. It was horrifying.

Mengele was a doctor who became mad because of the power he was given. Nobody ever questioned him—why did this one die? why did that one perish? The patients did not count.

He professed to do what he did in the name of science, but it was a madness on his part.

In this sham universe of scientific truths, half-truths, and outright lies, the experiments administered on the children represented a catalog of criminality and cruelty. Blood would be transfused from one twin to the other, and the results duly noted and compared. Bizarre psychological tests, designed to measure endurance, were continually made. For instance, a small child would be placed in isolation, in a small, cagelike room, with or without his twin. The children would be exposed to various stimuli and their reactions recorded. There are twins who recall they were targets of an insidious psychological barrage, but years later were still too traumatized to conjure up the details. And finally, there were the surgeries, the horrible, murderous operations.

Mengele would plunder a twin's body, sometimes removing organs and limbs. He injected the children with lethal germs, including typhus and tuberculosis, to see how quickly they succumbed to the diseases. Many became infected and died. He also attempted to change the sex of some twins. Female twins were sterilized; males were castrated. What was the point of these ghoulish experiments? No one, neither the child-victims nor the adult witnesses, ever really knew. Mengele constantly probed the children, trying to wrest from them secrets they did not possess.

VERA BLAU:

Mengele performed some very painful experiments on my sister, Rachel. She was very ill during her entire stay at the camp.

They would wheel her in and out of the operating room. But she does not remember what they did.

She remembers only the lights—the big red lights flashing down on her as they were about to operate. . . .

Several twins believe that Mengele had pairs of twins mate. There are hushed testimonies to that effect. Although all the twins deny firsthand knowledge, and many insist it never happened, there were rumors around the barracks that such perverse experiments were indeed taking place. No twin will elaborate on what he or she knew: Even in the nightmare world of Auschwitz, there were taboos, and this was the ultimate one. That Mengele breached it is not unlikely, given the awful scope of his experiments. We will probably never know for sure. Unless, of course, one twin, haunted by the memory of the forcible incestuous coupling, steps forward and testifies.

The final step in Mengele's scientific program was to kill the children and have their organs analyzed. Occasionally, though, if a twin seemed to be an especially "interesting" specimen, Mengele skipped the usual series of experiments and simply injected him or her with a shot of phenol to the heart. An autopsy would be performed, and various limbs and organs sent to Verschuer's Berlin institute. Mengele's mentor, in turn, subjected these to further scrutiny.

If Mengele's experiments seemed to each twin to become ever more diabolical, that was only an illusion: The intended objective had always been the death of the children. What the twins who had arrived in the spring and summer of 1944 did not know was that many other groups had preceded them. It was only a matter of time before their turn came.

MOSHE OFFER:

One day, my twin brother, Tibi, was taken away for some special experiments. Dr. Mengele had always been more interested in Tibi. I am not sure why—perhaps because he was the older twin.

Mengele made several operations on Tibi.

One surgery on his spine left my brother paralyzed. He could not walk anymore.

Then they took out his sexual organs.

After the fourth operation, I did not see Tibi anymore.

I cannot tell you how I felt. It is impossible to put into words how I felt.

They had taken away my father, my mother, my two older brothers—and now, my twin.

3

THE ANGEL OF DEATH

HEDVAH AND LEAH STERN:

There was a yard at Auschwitz, and there we waited for the airplanes. For the British. For the Americans. For the Russians. For anyone who could save us.

❧

MENASHE LORINCZI:

When we spotted the planes, we would pray, "Please God, bomb Auschwitz! Even if you have to kill us in the process, bomb the camp."

We were ready to die if it would mean the end of this horror.

But they never did. We watched the planes fly over us, but none of them ever dropped a single bomb on Auschwitz. I could not understand it—none of us could.

Every day, thousands of Hungarian Jews arrived on the trains. We watched them disembark. We saw them being herded into the crematoriums, which were working twenty-four hours a day.

❧

TWINS' FATHER:

The children would stand for hours just watching the flames.

The crematoriums were located only one hundred meters away from the twins' barracks. There was just a small fence in between.

The twins could see the transports—the trains pulling into Birkenau. They missed nothing that was going on at Auschwitz.

In front of the compound, there was a place where the Germans collected the dead bodies. There would be fifty, sixty, or a hundred corpses piled one on top of the other. These were then taken to the crematoriums in little wheelbarrows.

But there were times the people were still alive. And the children could see that some of these "corpses" were still living.

All day, the twins observed Mengele motioning people to go to the right and to the left. They watched the masses of people going into the crematoriums.

❧

MENASHE LORINCZI:

And even though we were children, we understood.

By the summer of 1944, Mengele's twins were among the very few people who had survived the selections at Auschwitz. Hardly anyone now arriving on the transports had the good fortune to be chosen for enslavement in labor camps. The war was now going progressively worse for the Nazis, and they stepped up their efforts to destroy their Jewish victims. Russian bombers regularly flew over the concentration camp, as did an occasional American plane. Hitler's army was surrounded by troops from both East and West, and there were reports of massive German defeats. With each new loss on the battlefront, the Auschwitz commandants took their fury out on the Jewish prisoners, herding them mercilessly into the gas chambers.

As Germany rushed to implement the Final Solution, more and more transports rolled into Auschwitz. Throughout that terrible summer, convoys pulled into nearby Birkenau with unrelenting regularity, delivering the final remnants of the Jews of Austria, Holland, Greece, Italy, France, Romania, Czechoslovakia, and, of course, Hungary.

The Hungarian Jews were the last intact Jewish community in Eastern Europe. Although formally allied with the Nazis, the Hungarian government had managed to shield its Jewish population for most of the war. But in 1944, the Germans occupied Hungary and assigned Adolf Eichmann, the architect of the Final Solution, to oversee the deportation of the Jews to the death camps. As if to make up for lost time, the Nazis sought to destroy the Hungarian Jews with a vengeance. By the summer of 1944, nearly half a million of them had been deported to Auschwitz. In order to accommodate the accelerated extermination process, four crematoriums functioned around the clock.

Mengele could be seen day and night at the head of the selection line, greeting the influx of new arrivals. Except for the twins, however, Mengele's finger pointed in only one direction now: toward the gas chambers.

PETER SOMOGYI:

My family arrived from Hungary on one of the last transports to Auschwitz—maybe the very last one.

We came from Pécs, a small town located several hundred kilometers from Budapest. We led a comfortable life before the war. My father was a representative for Ford Motor Company. He ran a business selling American cars, which were quite popular.

My older sister, my twin brother, and I had a nanny. All three of us attended a Jewish elementary school. Our family was very religious. There was a flourishing Jewish community in Pécs in those days.

Life was fine until 1944. The Nazis came to Pécs in March of that year, and everything changed. One anti-Semitic decree after another was issued. First, we had to wear a yellow Star of David. Then, we were no longer allowed to attend school.

In April, all the Jews of Pécs were rounded up and placed in a ghetto. We lived in this ghetto for about two months. Although it was very crowded, life was still not too bad. I can even remember a Boy Scout troop being formed inside the ghetto. We played regular games of soccer.

No one knew what was going on, what was going to happen. But many people—including my mother—suspected and feared the worst.

One day, the Nazis told us we were going to move. We would be relocated "somewhere in Austria." They were very vague. And they never, ever mentioned Poland—and certainly not Auschwitz.

We were taken out of the ghetto and placed in a large stable—a place where before the war our horses had been kept. The Germans also brought trainloads of Jews from other neighboring villages, and crammed them with us in this stable.

The first week in July, all the Jews were marched down to the train station and placed in cattle cars. We were packed in so tightly we could hardly breathe.

It took nearly four days to get to Auschwitz. Since this was the summer, it was extremely hot inside our car. We were given no food, no water. I remember a little boy crying incessantly for water.

We arrived at Auschwitz on July 9, 1944. It was early in the evening, and when we stepped out of the cattle cars, we could see the chimneys, with very, very high flames leaping out from them.

"What is this?" my brother and I asked our mother.

"Oh, it's probably just a big factory," she told us.

Uniformed guards walked up and down the selection line asking for twins. They asked for them both in German and Hungarian.

But it wasn't until the third time the guards asked that my mother admitted we were twins. The first two times, she kept silent. She didn't know what it meant, whether it was good or bad. And because my brother and I didn't look alike at all, it was easy for her to pretend we were not twins.

My brother and I were quickly plucked out of the line. It was the last time I ever saw my mother or my sister.

We were placed in an ambulance, and whisked off to the twins' barracks. There, we were greeted by Zvi Spiegel, who told us he was the "Twins' Father," in charge of all the boy twins.

We didn't know what had happened to our mother, and so we asked him when we could see her. He hedged. He didn't want to tell us yet what had happened, what was probably happening that very moment—that she had gone up in the flames.

Mengele's passion for selecting victims for the gas chambers, his cool efficiency and relish for the job, would earn him the sobriquet "the Angel of Death." It is unclear how the dandy from Günzburg first acquired the title. The preponderance of evidence suggests he got the nickname after the war, as the public gained awareness of his heinous deeds. Was it coined by an anguished survivor, haunted by the memory of Mengele's tender smile and ruthless acts? Or was it simply the product of some clever headline writer, invented when news of the Auschwitz doctor's crimes first surfaced? Whoever the author was, no Torah scholar or Hassidic grand rebbe, no mystic of the Cabala or Talmudic sage, could have envisioned a more perfect earthly incarnation of the evil spirit the Bible calls the *Malach Hamavet* than Dr. Josef Mengele of Auschwitz.

The Angel of Death is a figure who appears throughout the Old Testament. By chilling coincidence, the biblical lore even states he assumed the form of a physician, one, moreover, "of excellent repute." The ancient spiritual leader of Bratislava, Rabbi Nahman, once observed that "it was difficult for the Angel of Death to slay everyone, so he found doctors to assist him." Terrifying, utterly without mercy or compassion, the Angel of Death visited the earth clothed in a doctor's garb and cut an endless swath of destruction. "Even if the Almighty were to order me back upon earth to live my life all over again, I would refuse because of the horror of the Angel of Death," said Rabbi Nahman—a view that many of Mengele's victims would doubtless have shared.

The bizarre, almost poetic title stuck because it captured so well the contradictions in Mengele's character. Like the spirit *Malach Hamavet*, Mengele was a master destroyer, a Satanic figure brimming with evil and without regard for the value of a human life. But also like his namesake, Mengele was "angelic" in appearance and demeanor, able to charm, to woo, to captivate, to trick and seduce, everyone he met, most especially young children.

PETER SOMOGYI:

When we first met Dr. Mengele, my brother and I noticed he was whistling. Both of us had studied classical music in Hungary, and we recognized the tune as a work of Mozart. We told this to Mengele.

He was absolutely delighted. It made for an instant rapport. We could also speak German fluently, and Mengele seemed very pleased with that, as well.

We became Mengele's special protégés. He nicknamed us "the members of the intelligentsia."

Mengele related to the twins on different levels. With my brother and me, he liked to discuss music. We had long talks with him about culture. Perhaps because of this, we were not afraid of the experiments—or of him.

Once or twice, he took us to his office. There, he measured us, weighed us. He checked the size of our heads, of our eyes. He did this very gently.

I remember thinking Mengele was rather a nice man.

The massive number of Hungarian transports sent many more guinea pigs Mengele's way. Although the twins were from the start the focus of Mengele's work at Auschwitz, like every other SS doctor there he had his share of routine duties, such as signing death certificates and making sure outbreaks of contagious maladies like TB and cholera did not get out of hand.

Occasionally, to relieve the tedium, Mengele would demand a show of the new arrivals. One especially hot July day, a group of Hungarian rabbis descended from the cattle car. Despite the stifling heat, they were still wearing their traditional garb—long black coats, black woolen trousers, and fur hats. Mengele looked them up and down with contempt, then decided to amuse himself before motioning them to the gas chambers. First, he ordered the rabbis to step out of the selection line and sing. The holy men obeyed without a word of protest. Then, Mengele commanded them to dance. He wanted them to raise their arms and their voices toward the God who would not save them, no matter how loud their prayers.

And so the Hungarian rabbis began to dance, slowly, ponderously, under the sweltering sun of Auschwitz. They held their heads high, determined to preserve their dignity. With their eyes fixed to the sky, the rabbis chanted the "*Kol Nidre*," the mournful hymn of atonement, while Josef Mengele listened, unrepentant.

Mengele's zeal for his work clearly impressed his superiors, who showered him with accolades that terrible summer. He was now at the height of his powers. As chief doctor of Birkenau, the huge extermination center next to Auschwitz, he presided over a doomed population of Gypsies, twins, and several thousand female inmates. Although there were Nazis at Auschwitz who held higher ranks, none was as hated, or as feared, as Dr. Josef Mengele.

In the terrible summer of 1944, Birkenau was also crammed with hordes of newly arrived Hungarian women. Because of the stepped-up number of transports, even Nazi efficiency was proving inadequate to the task of slaughtering all the Jews who kept arriving. Each day, hundreds of Hungarian women were herded into Birkenau. There was simply no time to put these women through the normal selection process as they got off the trains, or even to assign them a number, and Mengele was obliged to perform the selections later, inside the women's camp.

Mengele was frequently accompanied during these inspections by a beautiful young German guard, Irma Grese. The "Blond Angel," as she was called, and the elegant Dr. Mengele made a splendid couple. Both were renowned around the camp for their beauty and sadism. Only eighteen, Irma loved to parade in her finery—clothes looted from the trunks of Jewish women. She gloated over the inmates who were forced by circumstance to be at her mercy. She cracked her leather whip on the helpless women like some Hollywood parody of a female SS guard. To Mengele, of course, she was utterly deferential.

The women of Birkenau both feared and admired Mengele. Much as they loathed to admit it, several of his female victims actually found him attractive. As he inspected them, some intuitively resorted to preening gestures of a time gone by, patting what was left of their hair, straightening their tattered camp uniforms, and attempting a smile. Clearly, many of these poor women were merely using their sexuality in a desperate effort to save themselves. But despite their physical frailty and emotional anguish, some were not immune to Mengele's sexual magnetism.

Mengele seemed as at ease with the adult women of Birkenau as he was with the twins, and all the more confident of his own attractiveness. To this day, survivors note the extreme care he took with his appearance—his uniform, exquisitely tailored and perfectly pressed; his cap, so carefully angled on the head; and his white gloves, which he wore even while making selections. Indeed, to these forlorn women, destined to die, Mengele seemed almost a romantic figure. Back in Günzburg, he had exuded charisma as he strode through town with his brisk, energetic walk and the slight smile that never left his face. There, he could have his pick of all the town belles. And the same was grotesquely true in the concentration camp. Here, too, Beppo Mengele could select any woman he pleased—as his next victim.

As part of the selections Mengele conducted inside Birkenau, women were required to undress and parade naked in front of him. This enabled him to judge whether they were fit to live just a little bit longer. Much as he was able to convince the young twins to like and trust him, Mengele intuitively knew the secret of making these female inmates feel at ease. Several of the women on parade would confide in him, admitting, in response to his questioning, that they were not feeling well, or suffered from some chronic ailment. What these women did not realize was that by trusting the handsome young doctor, and revealing to him their weaknesses and maladies, they were only hastening their own deaths.

JUDITH YAGUDAH:

When Mengele made the selections inside Birkenau, the women were not sent immediately to the gas chambers. They were first taken to another compound, not far from the twins' barracks.

These poor women knew they were going to be killed, and so they were constantly shouting and crying.

I would see big open trucks filled with naked women who were obviously destined for the gas chambers.

It was an awful, awful sight.

Once in a while, Mengele seemed to be drawn physically to one of the Jewish inmates, in spite of the efforts of their Nazi captors to strip them of their beauty. Shortly after arriving at Auschwitz, women were herded into Hitler's version of a "beauty parlor," where their heads were shaved. The women were then given either a regulation striped uniform or absurd, ill-fitting rags to wear. Their shoes were usually too large or too small, often consisting of a "pair" of one flat slipper and one high heel. The effect was to make the women look ridiculous and thoroughly unalluring to everyone—including their Aryan guards.

Occasionally, the Nazi system failed, and a woman radiated beauty, her shapeless garments and shorn hair notwithstanding. Mengele encountered such a woman in Ibi Hillman. Tall, blond, and statuesque, fifteen-year-old Ibi had been the pride of her small village in Transylvania. When Ibi removed her uniform in the course of an "inspection," Mengele found himself staring at her, transfixed. The other female inmates, and even his own assistants, watched him, aware of his attraction to the young Jewish woman. Any other SS officer would have simply made her his mistress. But Mengele evidently could not and would not concede feeling an attraction toward a Jew. In a loud voice, he dispatched Ibi to the infamous Block Ten, where the Nazis were performing sinister gynecological experiments. Few women survived Block Ten.

A few weeks later, Ibi was spotted by some inmates wandering by herself, in a daze. They hardly recognized her. The beautiful young girl looked like a shriveled old woman. Her slender limbs were now swollen and disfigured, while her stomach was bloated from the numerous surgeries that had been performed on her. Sickly and grotesque, Ibi Hillman no longer held any possible attraction—for Dr. Mengele or any other man.

Mengele also seemed to take a perverse pleasure in exterminating women who were pregnant. "This is not a maternity ward," he replied easily, when asked why these women were sometimes automatically sent to die. Mengele even boasted he was being "humanitarian" in having these women killed. Auschwitz, he would point out, had no facilities to take care of newborn children. But Mengele fluctuated erratically in his policy. Some weeks, he issued orders that pregnant women were to be kept alive and given every consideration. Other weeks, he ordered them killed immediately. At times, Mengele permitted a woman to deliver her baby, but then he promptly dispatched mother and infant to the gas chambers.

Even by Auschwitz standards, Mengele's obsessive cruelty to pregnant women stood out. When he first met an expectant inmate, Mengele liked to quiz her at length about her condition. He asked dozens of precise, detailed questions that allowed him to assume the protective coloration of a concerned physician. But the questions were often more personal than scientific, more voyeuristic than impartial. When did she become pregnant, he wanted to know, before or after arriving at Auschwitz? By whom? In what circumstances was the child conceived? In posing the questions, Mengele would try to maintain his usual detachment, but the intensity of his curiosity seemed odd both to the unfortunate women he addressed and to the assistants who overheard such exchanges. Pregnancy, after all, was an everyday occurrence. And unlike twins, triplets, dwarfs, or giants, pregnant women could hardly be deemed a scientific phenomenon.

MAGDA SPIEGEL:

Pregnant women were always coming to Mengele's office. He wanted to be present at the birth of their children—he wanted to be present during each and every birth at Auschwitz.

There were red-brick ovens in the middle of the barracks. The women were forced to give birth on these ovens: That was where Mengele "delivered" the babies. These poor women were given nothing, no pillows, no blankets.

❧

JUDITH YAGUDAH:

If a woman was pregnant, or even if Mengele simply thought she was pregnant, he sent her immediately to the gas chambers. My aunt, my mother's sister, was a bit overweight. She had a stomach. Mengele was convinced she was pregnant, and so he sent her to be gassed.

Mengele's Jewish assistants, like Dr. Gisella Perl, took to performing abortions just to save the lives of women who faced automatic death if their pregnancies were discovered. In an effort to rescue as many of these women as possible, Perl secretly performed crude abortions. When a pregnancy was too advanced, she would deliver the baby, then kill it with an injection of phenol, telling the mother her baby had been born dead.

Perl, who worked closely with Mengele, came to know him well and to despise him. In *I Was a Doctor in Auschwitz*, a searing account of her experiences at the concentration camp written shortly after the war, Perl described Mengele as a supreme sadist who lorded his power over the frailest subjects of Auschwitz's kingdom of Death: the pregnant women and their newborn children, the deformed hunchbacks and freakish giants, the dwarfs and midgets. Mengele was "so proud of his index finger which could distribute life or death at will, of his attractive, elegant physique . . . [of] his sham-medical profession!" Perl wrote with emotion. The people who caught Mengele's eye, she pointed out, were life's most vulnerable.

As chief doctor of Birkenau, Mengele was also in charge of the

Gypsy camp. Located near the twins' barracks, it was unique in the death camp, where the rule was to separate family members the moment they arrived. The Gypsies were allowed to stay together, perhaps because they were faithful Christians, despite their inferior racial stock. It was their one privilege.

Several thousand Gypsies were crowded into one encampment, whose borders were grimly defined by the crematoriums. At any time of the day or night, they could look up and see the blood-red smoke pouring out of the chimneys, or the Sonderkommandos pushing streams of people to their deaths. Because of the overcrowding, insufficient food, and poor sanitary conditions, epidemics ravaged the Gypsy population.

The wretched conditions notwithstanding, the Gypsies alone among the inmates of the concentration camp had the comfort of being with their loved ones. Their encampment looked like a vast playground. In an open area, dark-haired, dark-eyed men and women with traces of their former beauty sat watching their children at play. The youngsters ran around as if they were still in their own camps in the crossroads of Europe. It was an ongoing carnival, right in the middle of Auschwitz. The adults exchanged stories, sang songs, even danced. Some of the old melodies were so haunting that even the SS guards making their rounds would stop to listen. The music seemed to conjure up another life, far from Auschwitz and the war.

Mengele, the benevolent despot of this enclave, passed through the camp every day on the way to his laboratories. He appeared fond of these children, too, and they liked him in return, just like the twins in the compound next door. When Mengele came to see them, they crowded around him, extending their hands for candy, daring to reach for his bulging pockets that promised chocolates and other treats. "Uncle Mengele! Uncle Mengele!" they cried out. And Mengele would pat the children's heads and, reaching into his pockets, smilingly retrieve a bonbon or two. At the same time, he kept his eye out for potential experimental subjects. Tests were being conducted on several pairs of Gypsy twins, and Mengele was always watchful for more.

There was one little boy of exceptional beauty who was Mengele's favorite companion as he made his daily rounds at the Gypsy camp. What a striking pair they made—the tall, graceful doctor and the dark, delicate child who barely came up to his knee. Mengele had him dressed all in white, so that the boy looked strangely regal. Sometimes he would ask the little boy to perform a jig or sing a melody. Afterward,

Mengele would lean over and hug him, and ply him with chocolates and candies.

Mengele knew that conditions in the Gypsy camp were deteriorating quickly and that the death rate was mounting. He also knew why the Gypsies' situation was so abysmal: The Nazi hierarchy in Berlin had decided to exterminate them. Although Berlin kept wavering, it was only a matter of time before the inhabitants of the camp were to be put to death.

As their fate was being decided, the Gypsies formed little orchestras. They played tinny waltzes, lively mazurkas, touching ballads and operettas. Little girls danced to the music. The Gypsies always played when Mengele passed through because they knew how much he loved music.

The summer of 1944 dragged on. They were starving to death! Gypsy babies died, emaciated, after only a few days. There was no running water. Still, they danced. They did not realize it was a dance of death. Only Mengele knew.

The order finally came down from Berlin on August 2, 1944. At seven o'clock that evening, the Gypsy camp was sealed shut.

MENASHE LORINCZI:

We heard a terrible cry. The Gypsies knew they were going to be put to death, and they cried all night.

They had been at Auschwitz a long time. They had seen the Jews arriving at the ramps, had watched the selections where old people and children went to the gas chambers. [And so] *they cried.*

And when the Gypsies cried, all the twins heard them.

And even though I was a child, only nine or ten years old, I understood.

Toward the end of that terrible night, when nearly all the Gypsies had been slaughtered, Mengele went to get the boy who had been his little mascot. Hand in hand, they walked around the camp, as they had done for so many months. Then Mengele took the child to the gas chamber and showed him the way inside. He obediently climbed in.

Efficient as ever, Mengele did not forget the sets of Gypsy twins on whom experiments were being conducted. They were gassed with the rest of the Gypsies, but he had the letter X marked on their chests so they would not be cremated. The corpses were duly delivered to his pathology laboratory, where his assistant, Dr. Nyiszli, performed analyses of them as ordered. His work complete, Nyiszli turned over his reports to Mengele. That same night, the two sat for hours calmly discussing the findings.

A few weeks later, in September, Mengele's wife, Irene, arrived. It was her second trip to the death camp. She had been there the previous year as well, and had brought along little Rolf. This time, she left the child behind in Germany.

Irene and Josef had spent little time together since their marriage four years earlier. Although Mengele's wife was certainly aware of what was happening at Auschwitz, her diary suggests she did not let it mar her reunion with her husband. The couple went swimming and enjoyed outings in nearby fields, where they picked berries. Irene's Auschwitz holiday was interrupted when she contracted typhoid fever and had to be hospitalized. After receiving every care, she became well enough to travel back to Germany with Josef. In an entry she made after the trip, she observed that her darling husband had seemed a bit depressed. She did not elaborate. But the course of the war—as well as his activities in the camp—may well have been weighing on Mengele. On a visit made to the Berlin home of Professor Verschuer at about this time, Mengele appeared dour when asked how Auschwitz was. "Simply horrible," he replied.

In those final months of the war, Mengele made more selections than ever. To observers, he acted like a man possessed, as though he was driven by the notion that if he worked hard enough, performed enough experiments, tested a sufficient number of twins, he might still make some important scientific discovery.

One of Mengele's greatest fears was that the hated "Bolsheviks" would get their hands on his research material, so painstakingly collected over the last couple of years. One day, he invited Dr. Ella Lingens, a non-Jewish physician who was a political prisoner, to review the reports he had compiled. As Lingens leafed through the thick sheaves of papers, charts, and graphs, Mengele bitterly observed, "Isn't it a pity all this work will fall into the hands of the Bolsheviks?"

And, indeed, that fall Russian troops were reported to be rapidly approaching. As they drew nearer and confusion at the camp mounted, the twins' protected status became increasingly imperiled.

PETER SOMOGYI:

One day, a new doctor came around to inspect the twins. His name was Dr. Thilo. He made a selection, which had never been done on us. He selected all the male twins—we were all to be sent to the gas chambers. Every twin's number was marked down as destined for the crematorium.

Our barracks were sealed. All the doors were boarded up. We were not allowed to leave.

❧

ALEX DEKEL:

It was November when I was sent into the barracks with the other children. I was thirteen at the time, and I knew exactly what was going on. I knew everyone was selected to go to the gas chambers. You were selected in the afternoon, and by midnight you were picked up and sent to the gas chamber.

Our barracks had small windows—they were four feet, five feet, off the ground. I jumped up to one of those small windows. It was impossible to reach, but I jumped. Somehow I reached it, and I just hung there, barely grasping the windowsill with my fingers. Then, I do not know how, I managed to jump outside. I landed in the snow and saw a building in front of me. I climbed through the window and found myself in a toilet. Outside, I heard truck engines. And so I jumped down into the toilet. I heard a German guard enter. He took a flashlight, shined it around the room, but he didn't see me. I stayed there all night, hidden inside the toilet.

❧

TWINS' FATHER:

The children were very upset, very frightened. They knew what was going to happen to them. They knew the end was coming—that the Nazis were planning to kill them.

❧

PETER SOMOGYI:

I remember trying to plot ways to get revenge. I had a little pocketknife—we all did, to cut our bread in the morning. The Nazis never thought there was any danger in giving little children pocket-knives.

I remember sharpening my knife like a fanatic. I knew we would be taken to the gas chambers in trucks, with several SS guards inside to watch over us.

Although I was only eleven years old, I remember this very distinctly, wanting to kill a Nazi. I said to myself, "Before I go, I am taking at least one SS man with me."

❧

TWINS' FATHER:

I don't know how I got the authority, but somehow I was able to leave the barracks.

I ran toward Mengele's office. I ran even though it was very dangerous to run through the camp, because the SS shot anyone who moved too quickly. But I knew time was of the essence.

I told the guards outside Mengele's office that I wanted to speak with him. This was a bit like saying you wanted to speak with God. To this day, I don't know why they didn't shoot me for that request, either. But somehow, I was allowed in to see Mengele. I told him about Thilo's order to kill all the twins.

Mengele was quite upset. He immediately went to reverse the order, and said the children should be kept alive.

❧

PETER SOMOGYI:

The Nazis came and opened up the doors to our barracks. To our surprise, we were told we could go outside.

When Twins' Father returned a bit later, we kept asking him, "What happened? Why didn't we die?"

He told us simply that Mengele had called off the selection.

Later that fall, the twins were alarmed to hear they were being moved into the deserted Gypsy barracks. No explanation was given for the change. The children were afraid they had been marked again for extermination. Rumors spread through their barracks, despite the fact that Mengele was proceeding with his tests and experiments as diligently as before. Even with the end closing in on him, even as Hitler, his generals, and his professors brought Germany to ruin with their notions of a superior race, Mengele still hoped to make some great scientific discovery that would validate them all, save the crumbling Third Reich—and his own disintegrating dreams.

JUDITH YAGUDAH:

I still remember when they moved us into the Gypsy camp. It was very cold—below zero—and it was snowing.

❧

MENASHE LORINCZI:

We thought it was the end of us.

❧

JUDITH YAGUDAH:

They made us stand outside because a prisoner was missing. We stood for hours in the cold, in the snow, until they found this prisoner.

But as a result, my sister Ruthie got severe frostbite. Her feet were frozen. She had to have an operation, and they removed some of her toes.

She could not walk after that.

Back home, in Hungary, Ruthie had always been the livelier twin. She had loved to dance.

Ruthie kept asking my mother if we would ever go home, and if she would ever dance again.

That was her main concern—whether she would dance again.

4

THE ANGEL VANISHES

For the inmates of Auschwitz, December 31, 1944, marked the long-awaited time of hope and celebration. The camp was quiet that New Year's Eve. Many of the SS guards and commandants were no longer there, having fled to avoid being captured. The Russian Army was now just a few kilometers away.

At previous New Year's festivities, bedraggled inmates of the death-camp orchestra had played through the night for the Nazis. Prisoners would lie awake listening to the drunken revelry. But this year, the only music to be heard was the soft whir of Russian planes circling overhead. Air-raid sirens sounded intermittently, plaintive and insistent.

A furtive party was held in Dr. Mengele's infirmary by a group of nurses and doctors who were also prisoners. The celebrants traded news about the fall of Berlin and the imminent Russian arrival; they toasted the future by clinking cups of soup.

Auschwitz was no longer the formidable death camp it had been. Those Nazis who remained on duty through New Year's were busy trying to cover their tracks. They scurried around destroying documents and photographs, attempting to obliterate any evidence that would reveal the enormity of their crimes. Many of the crematoriums had already been dismantled, the culmination of a process undertaken in November and December on direct orders from the Nazi hierarchy in Berlin.

Several of the warehouses containing the personal belongings of the Jews had been emptied of their goods. Between December and January, for instance, 514,843 articles of men's, women's, and children's clothing were shipped out, according to an official Nazi report. But the camp where four million people had passed through—of whom only sixty thousand survived—had more evidence than even the Nazis' able hands could destroy. Entire storerooms remained intact, crammed from floor to ceiling with the belongings of the millions.

One morning in late December, Mengele had marched into his laboratory and announced it was being moved to another crematorium. The equipment was carefully packed away and transported to the one crematorium that was still operational. The German Army was in shambles and the Russians just weeks away from storming the camp, yet Mengele ordered his assistants to set up the dissecting tables exactly as before, in preparation for more "work."

Although Mengele was maintaining his usual studied nonchalance to the last, he was secretly planning his own escape. But unlike the other camp doctors, who destroyed their work, Mengele didn't want the results from a single experiment to be destroyed, or worse, to fall into the hands of the despised Russians. Mengele apparently took steps to safeguard his work. Key slides and specimens were carefully packed and sent to Günzburg for safekeeping. He could only hope that Professor Verschuer in Berlin-Dahlem would find some way to preserve the hundreds of documents, reports, tissue samples, and organs he had sent him over the last eighteen months.

Late one night, sometime in the middle of January—no one is certain exactly when—the Angel of Death vanished. He had betrayed no hint he would be leaving, even as he visited the twins and inspected the surviving women of Birkenau for what he knew would be the last time. One prisoner did notice him putting boxloads of documents into a waiting car. He left the camp so stealthily that for days even some of his assistants failed to realize Mengele was gone. He was one of the last SS doctors to leave Auschwitz.

The remaining SS command was torn between a desire to destroy all remaining evidence, which would take time, and the more primitive mandate of *Sauve qui peut*—getting away while they still could. In just a few days, the SS managed to blow up the most damning evidence of all: the crematoriums that were still standing. Explosions ripped the ovens that had consumed so many millions, until all that was left of them were heaps of ashes and piles of smoking rubble. The Nazis also set fire to the "Kanada," the storage room filled with so many treasures that the inmates had named it after that faraway country whose very name conjured up visions of wealth.

JUDITH YAGUDAH:

The whole camp was burning.

The Germans placed explosives under the barracks and storerooms. They were setting fire to different parts of the camp.

We thought the Germans were going to burn us alive. We ran out of our barracks. The sky was red with flames. It looked like an inferno.

It was a terrible period. The Germans were leaving the camp, and taking with them anyone who was able to go.

But my sister, Ruthie, couldn't walk. She was very sick.

My mother decided then and there we were going to leave the camp. She got a wheelbarrow—God knows from where—and lined it with blankets and pillows. She stuck Ruthie in the wheelbarrow, took me by the hand, and started walking. She was prepared to wheel Ruthie out of Auschwitz.

But a woman saw the three of us walking together and told my mother, "Are you crazy? Where are you going to go with your small children?" She told her to go back to the barracks and stay there.

My mother went to some of the storerooms and grabbed things for us. She took sweaters—coats—blankets—from among the beautiful items the Jews had brought. The Nazis had kept them in excellent order.

Eager to avoid the oncoming Russians, any guards still left at the camp had rounded up most of the remaining adult prisoners and marched them off into the frigid Polish winter to destinations unknown. Many of the twins were forced to join this Death March, as it later came to be called. With defeat finally imminent, the Nazi guards were even more brutal toward the enfeebled veterans of the concentration camp than ever before.

Vera Blau:

Anyone who stopped was shot instantly.

My sister, Rachel, kept saying she couldn't go on, that she had no strength to walk anymore. And so I decided to drag her along. I grabbed her by her coat and dragged her through the snow.

For four days, we walked in the snow. It was extremely cold. Once, we were given hot water to drink. I spilled the hot water, and it fell into my shoe. It was so cold, my foot got frozen.

Leah Stern:

The Nazis gave us neither food nor drink during this march. We were so thirsty, we wanted to swallow some of the snow. But the Germans wouldn't even let us do that.

The snow came to our knees. My sister, Hedvah, and I were wearing very thin clothes. We huddled under a blanket as we trudged through the snow. I was so weak, I kept wanting to stop, but my twin wouldn't let me.

She decided we had to lighten our load to make walking easier. First, we threw away the blanket. We even tossed out some bread we had taken with us. For weeks prior to leaving Auschwitz, we'd been saving this bread. But even those little crusts of stale bread were proving to be too heavy for us to carry.

When it looked like I was going to collapse. Hedvah hoisted me on her shoulder and carried me. She saved my life.

Alex Dekel:

We were forced to march in the freezing cold from Poland, through Czechoslovakia, to Austria and the Mauthausen concentration camp.

Every step I walked, I would sink to my knees and pray to God to kill me. I would get up, walk, and fall again, and again ask for death.

Zvi the Sailor:

We started the Death March with about twenty thousand people. After two or three weeks, we arrived at the Mauthausen concentration camp with three thousand people—maybe only two thousand.

On this march, the Nazis treated us worse than animals. They did whatever they wanted with us.

I recall marching across big fields, and hearing shooting. To this day, I can still hear the sound of these gunshots. Since that time, I have fought in many wars, and I have heard many different kinds of shootings, but I still remember this shooting more clearly than any other.

We kept marching and hearing these shots in the background. We came to a big river between two hills, and we were told to run. And so we ran. There was only one soldier, with only one machine gun, and he was shooting at us as we ran.

And today, I live with this memory, and I am ashamed. I ask myself, how could we have been so stupid? Because today I realize there was only one SS man doing the shooting, and hundreds—maybe thousands—of people who were running, terrified of this one, solitary SS man.

The twins who evaded the Death March remained at the largely deserted concentration camp. There was neither food nor water. With no guards in sight, those still at Auschwitz plundered any storerooms still left standing. Hoping that the rumors the Russians were on their way were true, Mengele's children waited to be rescued.

PETER SOMOGYI:

We were terrified the Nazis would come back. Even without German soldiers, there was no sense of security.

We were very hungry. There was no food, no food at all.

One morning, I decided to go out and hunt for food. I walked around the deserted camp. But all I found was a big warehouse filled with cases of bottled water. And so I picked up some bottles and brought them back to the twins.

Several of the male twins were together, and our Twins' Father was still with us, watching over us. He wanted to leave immediately and go east. He, too, was afraid the Nazis would return and take us prisoner—and we'd have no second chance.

Then one day we looked up to see the first Russian troops. . . .

On January 27, 1945, at three o'clock in the afternoon, Russian soldiers marched into Auschwitz. They found the twins huddled inside one of the barracks. They were cold and hungry. Many were suffering from typhoid fever and dysentery. Their frail, emaciated bodies still bore needle marks from the blood tests and injections that had been administered up to the end.

But they were alive! Mengele's twins were among the only children to have survived Auschwitz.

The first thing the Russians did was to distribute clothing and blankets. The twins were given the camp's striped uniforms, which they had never had to wear before. The jackets and trousers were so big the children had to stuff layer after layer of clothing to keep them from falling off.

That night, there was a big party to celebrate the liberation. The emaciated women of Birkenau discovered long-lost sources of energy within their frail bodies. In their joy, they danced all night with the Russian soldiers. Mengele's twins stood by happily, watching the adult goings-on.

The next day, the dazed children were marched out of the camp. The scene was captured by Russian cameramen as part of a propaganda film that would show the world how the Russian Army had rescued Jewish children from Fascist hands. The twins had to roll up their sleeves and show their tatooed numbers for close-up shots. They were led out of Auschwitz several times, until the director was satisfied the liberation scene looked sufficiently authentic.

As they passed through the gates of Auschwitz, twins clutched each other tightly. Amid the euphoria of the liberation was the aching sadness felt by all the children, knowing they were leaving Auschwitz without the families with whom they had arrived. The camp lay behind them like a vast cemetery, filled with the unmarked graves of their loved ones.

Journalists descended on Auschwitz in the immediate aftermath of the liberation, demanding to get the story that had eluded them for so many years: the story of the Nazis' systematic slaughter of millions of Jews in one cold and forsaken corner of Poland. Menashe Lorinczi, a twin who had worked as Mengele's messenger boy, volunteered to be their tour guide. Menashe took both reporters and Russian soldiers through the maze of buildings and barracks that made up Auschwitz-Birkenau, pointing out the various facilities. To his pride and delight, he became an instant hero, the first child survivor to be interviewed and quoted in newspapers around the world.

MENASHE LORINCZI:

Right after talking with the journalists and touring Auschwitz with them, I collapsed. The Russians put me in an infirmary they had set up at the camp.

I lay in bed for months, suffering from one illness after the other. My teeth fell out. I developed a lung infection, then tuberculosis.

Months after the liberation of Auschwitz, I was still sick, still in the concentration camp, with my twin sister at my side.

Some of the twins who were well and did not require hospitalization were bundled out of Auschwitz and taken to a large monastery in the Polish city of Katowice. There, for the first time in years, they were given plenty of food, clothing, and even toys. Best of all, they were free to roam the city. There were joyful rides on the streetcar, where they merely had to show the tattooed number on their arm for the driver to let them on without exacting a fare. At the monastery, the only discipline the nuns imposed on the children was requiring them to come back in time for meals, which were served promptly three times a day.

EVA MOZES:

I will never forget that monastery. The first night, Miriam and I were put in a beautiful room with a large bed, covered by the whitest sheets I had ever seen. The room was filled with toys. I had no idea what to do with all those toys. I felt so filthy and dirty—no one had bothered to give us a bath, you see. I had lice crawling all over me.

And so, that first night, I removed the sheet from the bed and fell asleep on the bare mattress.

Zvi Spiegel, who at thirty was one of the oldest of Mengele's twins, the one they had lovingly called "Twins' Father," was reminded of a promise he had made during the dark days of the mass exterminations. He had sworn to take all the boys home.

TWINS' FATHER:

The twins were very excited. They were like little bees buzzing around me. They kept saying, "You promised you would take us home, Twins' Father. You promised." It was true—I had promised this. But it had only been to make them happy. I had never believed it myself.

One of the older prisoners wanted me to run away with him. "We're free—let's get out of here together," he kept telling me. But I decided that I had to take the twins home. I felt an obligation toward them.

I left Auschwitz on January 28, 1945, with thirty-six children in tow. I made a list of all their names, their ages, and where they were from.

Before we started the journey, I gathered the children for a lecture. I laid down the rules, just as if they were troops I might have commanded in former years in the Czech Army. I told them they had to stick together and keep up with me, or else I couldn't be responsible for them.

❧

PETER SOMOGYI:

We begged Twins' Father to take us with him because we were so fearful of remaining at Auschwitz. We knew we were free, but we were constantly afraid the Nazis would somehow make their way back to the camp.

"What if the Germans return and kill us?" we kept wondering. Even with the Russian Army there, feeding us, taking care of us, we still worried about the Germans.

With Twins' Father leading the way, we left Auschwitz and started marching east, toward Kraków.

It was bitter cold. My twin brother and I carried these little knapsacks I had sewn from a blanket. In mine, I carried all my worldly belongings: a small crust of bread and some of the bottles of water I had found in the deserted storeroom. When I was "packing" to leave Auschwitz, I found I had nothing to take with me—no clothes, no food other than the little piece of bread—so I stuck these bottles of water in my knapsack. I figured they might come in handy.

I remember, as we walked, feeling icicles in my pants and legs. It was so cold that the bottles had exploded, and the water had frozen around my body.

We walked and walked—but we didn't allow ourselves to stop. We were very afraid of turning back. And even though we were freezing, Twins' Father would tell us to keep going.

❧

TWINS' FATHER:

It took us three days just to get to Kraków. On the way, we passed soup kitchens set up to feed Russian soldiers. We would beg the Russians to let us eat their food. Sometimes they'd say yes. Other times, they'd shoo us away, and we had to continue walking.

Along the way, we kept picking up camp survivors. First, we ran into a group of twelve women. They were from Birkenau and were trying to get home. These women thought that by traveling with a group of children, they might, somehow, find their own kids. They kept hoping other child survivors would join us—perhaps their own.

I took down their first and last names, their date of birth, their hometowns, even their barracks number at Auschwitz, and added them to my list. I gave these women the same order I'd given the twins: to stick together and keep up with the group. By the time we got to Kraków, an additional one hundred male survivors had also joined our group.

I felt as if I were leading a small army: 153 men, women, and children, to be precise. I know the exact number because I continued to keep a neat list. I am not sure why preparing those lists was so important to me in the middle of all the chaos and confusion. I suppose it was my own way of maintaining some form of control. Even at the camp, I was obsessed with maintaining lists and keeping the children in order. I felt it was the key to our survival. And I have kept those lists even to this day.

My system helped convince Russian authorities to let us through the various roadblocks. Wherever we stopped, Russian Army officials would interrogate me as to who I was and what I was doing. I'd show them the lists, which impressed them. They were generally understanding, and gave me official documents authorizing me to lead the twins home.

❧

PETER SOMOGYI:

On the last leg of the trip between Auschwitz and Kraków, a Russian truck picked us up and gave us a ride. When we got to Kraków, we found an abandoned house—it had no furniture, no heat, nothing. All the twins slept together on the floor.

Twins' Father found a Red Cross center that would feed us once a day. For the rest of our meals, we had to beg strangers for food.

I remember going around Kraców, marching from house to house every day asking for a slice of bread. Some of the Poles were nice to the twins. But many were very mean. Even though they knew we were survivors from the death camp, they told us to go away.

We got very little from the Russians. There were no provisions for people other than those in the Russian Army.

At one point, Twins' Father tried to get the Russians to give us a vehicle so he could take us home more quickly, but he couldn't arrange it. I guess they had other problems on their minds than the fate of a bunch of little Jewish war orphans.

❧

TWINS' FATHER:

The trip home was a nightmare. All of Europe was in disarray. The rail system had been completely destroyed. We would board a train, then another train, and get absolutely nowhere. At times, we would take a train, and it would take us in the opposite direction from where we wanted to go.

When we got to the Hungarian border, I decided I needed a system to make sure everyone could get home as quickly as possible. I was most concerned about the children, of course. I decided to divide them into little groups, splitting everyone up by where they were going.

I assigned older twins to be in charge of younger ones. I gave the senior boys my address—I assumed I was returning to my old house—and told them to let me know how the trip went. Then, I set out with a small group of my own toward Munkács, my native village in Czechoslovakia.

❧

PETER SOMOGYI:

My brother and I and a few other twins boarded a train bound for Budapest. Before hugging us good-bye, Twins' Father made sure we were going in the right direction.

There was no regular source of food, so once again we were forced to beg. We asked Russian soldiers on the trains to give us some bread.

We finally arrived in a small town inside Hungary, not far from Budapest. There were Jews living there who had somehow survived the war. They invited us into their home, and gave us some very greasy chicken soup. These Jews were very kind to us. But it had been so long since we'd eaten a real meal, we got very sick. We were ill for days, unable to move.

It took us a few more days to get to Budapest. Someone at the train station—I am not sure who—directed us to a Jewish orphanage. We were the very first survivors of a death camp to come back to Hungary, and our arrival made big news in the Budapest Jewish community.

Although the Hungarian Jewish community had been decimated by the Nazis in the last months of the war, many Jews in the capital had survived without being deported. Budapest hadn't been emptied of its Jewish population, as was the case with smaller Hungarian cities and towns. A lot of them had remained in hiding. Several Jewish families had even managed to remain intact.

One of my mother's cousins heard about us. He came to get us, and together we began the journey back to our hometown of Pécs.

Back at Auschwitz, Mengele's experiments claimed their last victim. For much of her stay at the camp, Ruthie Rosenbaum had been desperately ill. She had remained alive only because of the relentless efforts of her twin sister, Judith, and their mother. By the time the Russians arrived, she was dying. Hoping to save her, Russian doctors quickly administered powerful medications. But despite their efforts, she died on March 3, 1945, her six-year-old body worn out from the tests and fear, hunger, and pain—and the Russian drugs, which ultimately proved to be lethal.

Mrs. Rosenbaum and Judith began the long journey to their native Cluj, in Romania. The trip through devastated Eastern Europe, as thousands of other survivors were groping their way home, took months. But mother and daughter both hoped for a happy reunion.

JUDITH YAGUDAH:

When we got to Cluj, there was a list at the train station of the Jews who had returned. From our entire family, only one uncle had made it back. He had learned we were coming, and that only one twin had survived. But he didn't know which one—me or Ruthie.

His first question to my poor mother was, "Who came back, Judith or Ruthie?"

We learned from him that most of our family had been killed in the war, that most of the Jews of Cluj had died.

❧

PETER SOMOGYI:

We came home only to find there was no home. We quickly learned that not a single Jew from our town had returned: They had all died in the concentration camps.

My family's house had been vandalized, our furniture burnt by townspeople or soldiers to keep warm. But I had seen so much destruction that I took the destruction of my own home very matter-of-factly. During our march out of Auschwitz with Twins' Father, we had gone through a town where there wasn't a single house still standing. When you see an entire town in ruins, what can you say to the fact that your own little house is no more?

❧

TWINS' FATHER:

When I finally arrived in Munkács, I went straight to my old house. But I found there were strangers living there. The new owners were scared to see me. You see, they'd simply taken it over after the family was deported. They went out of their way to treat me nicely. They even gave me mail I'd received. There were letters from several of the twins, telling me they'd gotten home safely. There were even "reports" on how they'd accompanied their younger charges without any problems.

Inside the house, I didn't see very much of our old furniture. Only a big mirror we used to own.

It was awful—awful—simply awful.

I stayed in Munkács exactly one day and one night. Then I ran away. I couldn't bear to live in my own village anymore.

The twins who were still at Auschwitz under the Russians' care or in the monastery at Katowice had only one desire—to get home as quickly as possible and meet up with relatives. Some children set off alone. Others found adults willing to accompany them for part of the journey. For most, the long-awaited return home ended in bitter disappointment. Nearly all were orphans, many the sole survivors of their entire families, even of their whole village.

EVA MOZES:

For almost a year, we had lived for the moment when we would go home.

At the monastery, we were told we could be taken to Palestine. But my sister, Miriam, and I wanted to go back to Romania and see who had survived. We thought we would get home and there would be someone to greet us—that someone besides us must have survived the war.

Because of our adventures riding the streetcars, we knew of a displaced persons' camp in Katowice where several people we had known at Auschwitz were staying. There, we found the mother of a pair of twin girls who was from our hometown.

She agreed to come with us to the monastery and sign papers for our release. Together with her daughters, we began the journey home.

As we approached our old house, I was still hoping against hope that someone would be there. We reached the gate. The house looked neglected. There were tall weeds around it. It did not look at all as I remembered it.

Only our old dog, Lilly, was still there. It was the only familiar face. I guess the Germans had not deported Jewish dogs.

It was so different from what we had expected our homecoming to be. I was heartbroken. Miriam and I started running. We were both crying hysterically.

By the time Mengele's twins were liberated, the SS doctor himself had made his way to another concentration camp, Gross Rosen, in Poland's Upper Silesia. Several hundred kilometers from Auschwitz, it promised a temporary reprieve while the Russians advanced. There, Mengele slipped easily into the familiar routine of camp physician. At thirty-four, he was a veteran death-camp hand. He made inspection tours of the camp and signed death certificates. Alas, his duties at Gross Rosen included neither experiments nor selections. When it became known the Russians were close to overrunning Gross Rosen, Mengele prepared to vanish again. He left Gross Rosen just before the Russians reached it on February 11, 1945. After Gross Rosen, Mengele's route is unclear. However, he was spotted by some of his twins in yet another death camp, Mauthausen, in the no-man's-land of Czechoslovakia.

Mauthausen was situated in what was essentially the last German front. Several of the twins had ended up there after the Death March. At least two claim to have recognized "Uncle Mengele" immediately. If he saw them, or remembered them, he gave no sign of recognition.

MOSHE OFFER:

I was in Mauthausen for about two or three weeks after the Death March. Then I was put along with other camp survivors on a train bound for I don't know where.

It was very crowded. I felt a push, and I fell off the train. To this day, I don't know if it was because I was so skinny, or because someone was trying to save me. In any case, I landed in a field. My hand was broken from the fall from the train.

I crawled through the field, and I looked up to see a big German in uniform. I started crying and asked him to kill me. I told the Nazi I was a Jew. "I fell off the train, and I cannot take it anymore, so go ahead and kill me."

But the old German soldier said to me, "I won't kill you—I am going to hide you." And so he took me to an attic where they were storing corn and other grain.

This old German soldier was very nice. Every day, he brought me some dry biscuits and water. From the window of the attic, I saw the war coming to an end. I saw trains go by, carrying munitions. I watched German airplanes being shot down and parachutists jumping out—and being shot themselves.

One day, in the middle of the night, I heard artillery fire. After that night, the German didn't come anymore. I no longer got food or water. And so I began eating the corn and grain.

For four days, I was without any food or water. I was very hungry, very thirsty. But I stayed in the attic because I was also very frightened.

Then one day I looked out from the window and saw a jeep. It was carrying American soldiers. I was so weak, I couldn't walk, and so I crawled on my hands and knees from the attic to the jeep.

The American GIs spotted me and rescued me. They carried me in their arms to their jeep, and they gave me candy and chocolates. But I was so sick, I couldn't even eat them.

The GIs took me to one of their doctors near Linz, in Austria, who treated me. Then they gave me a little uniform to wear—an American GI uniform.

I became their mascot. I would stand in line to get food, just like the other soldiers. But they let me go to the head of the line. They gave me dollars, toys, suitcases filled with candy.

I told them what had happened to me, how I had been at Auschwitz, and that my entire family had been killed. Some American soldiers in the group offered to take me home with them to America. But I got very sick, and they had to leave me behind. I came down with a high fever, and I was diagnosed as having typhoid. I was put in a convent hospital in Linz, where for four months I lay in bed. I couldn't even open my eyes.

I never saw the American soldiers again.

I told the people in the convent I wanted to go home—to Hungary.

When Mauthausen seemed in danger of falling to the Russians in May 1945, Mengele quickly shed his SS uniform and donned the outfit of a soldier of the Wehrmacht. He joined a German field hospital, submerging himself in the sea of embittered soldiers of the defeated army. The unit wandered between U.S.-and Soviet-held territories. At last, the unit surrendered to the American Army, and with his comrades, Mengele became a prisoner of war.

HEDVAH AND LEAH STERN:

We were liberated much later than the other twins. In fact, when we first heard the war was over—we didn't believe it.

After the Death March, we had ended up in a small town in Germany called Pritzberg, and everything was very chaotic, very confused. Someone shouted, "The war is over," and we thought they were joking. When we realized it was true, we didn't know where to go, what to do. We wandered through the streets of Pritzberg, crying, in our torn, dirty dresses.

Some French soldiers in the town noticed us. They felt sorry for us—two fourteen-year-old girls, obviously war orphans. They picked us up in their jeep and brought us to their headquarters. There, we were fed—our first real meal since before the war.

It was hard to communicate with the French soldiers. We spoke no French or German—only Hungarian. But they were very nice. We were able to make them understand how much we wanted to go home—to our hometown in Hungary. They gave us two suitcases and some cans of sardines, and they put us on a train bound for Hungary. But instead of taking us home, it went in the opposite direction, to Czechoslovakia. At last, we managed to board a train to our hometown. There, we were met by two widowed uncles—the only members of our family who had survived.

We told them we were left alive because we were twins.

They took us by our old house. Everything had been taken—there was nothing left that had belonged to our family.

It was so sad. We couldn't stop crying.

Confined to a POW camp near Munich, Mengele became deeply depressed and even contemplated suicide, according to his written account of that period. He confided his situation to a fellow prisoner, a doctor named Fritz Ulmann. Ulmann had an extra set of ID papers, which Mengele gratefully accepted: They would be extremely useful should he need to conceal his true identity. He carefully altered the card and became Fritz Hollman. By June 1945, American Occupation forces had begun rounding up Nazi functionaries and throwing them in jails and internment camps. More than fifty-thousand suspected Nazis—Gestapo heads, Hitler Youth leaders, members of the Nazi Peasants' League, SS officers—were imprisoned.

Mengele would have been a prime candidate for automatic arrest had the Americans been able to identify him. They believed they had developed a foolproof technique for singling out former members of the SS. Since all SS officers had been required to have their blood group tattooed beneath one arm, American soldiers looked for the tattoo itself or for a sign that it had been removed. As it turned out, Mengele's supreme vanity saved him: When he had joined the Waffen SS in 1940, he had refused on aesthetic grounds to submit to the tattoo. On August 18, 1945, Mengele was released from the POW camp by soldiers who saw no evidence that he had been in the SS.

According to his son, Rolf, Mengele returned to his parents' home in Günzburg sometime during that summer. He did not stay with them because it was too dangerous; instead, he was forced to hide out in the woods, where his family furtively provided him with food. But U.S. Occupation forces were establishing outposts around Germany, and Mengele was compelled to seek another hiding place, according to an autobiographical novel he wrote years after the war. The "novel" was among the papers Rolf turned over to the German magazine *Bunte* in 1985. *Bunte* obligingly gave it to the German prosecutors, who passed it along to the United States Justice Department. Although purportedly a fictional account of the adventures of a World War II veteran, the book so closely dovetails with actual events in Mengele's postwar life that it was used by the Justice Department in piecing together the war criminal's path after leaving Auschwitz. Mengele evidently decided to contact Ulmann's brother-in-law, a doctor practicing near Munich (he is described pseudonymously in the novel). This doctor sympathized with the war criminal's plight, and provided Mengele with the name of a farmer who could provide him with temporary refuge.

Mangolding, a small farming community outside Munich, had not been touched by the war. When Mengele arrived at the farm of George Fischer, he told the farmer he was a refugee in need of a job. In an interview with *Bunte* magazine in 1985, Fischer revealed that he agreed to take Mengele on as a farmhand if the young man proved he could do the job. His first task was to sort the potato crop, separating potatoes suitable for human consumption from those of inferior quality, which could be used as animal feed. He was to make two piles: good potatoes to the right and bad ones to the left. Fischer warned Mengele to choose carefully; with the scarcity of food, potatoes were a precious commodity.

Mengele concentrated on the task at hand, and exerted himself to do it well—taking as much care, perhaps, as when he had served on the selection ramp of Auschwitz. He distinguished himself at potato-culling and got the job. As a farmhand, not only was he safe from the authorities, but in the midst of the confusion, starvation, and poverty engulfing the rest of Germany, Mengele was assured a good home and plenty of food. He even had time to devote to his favorite activity, reading. He read voraciously, staying up late to finish a book, then rising early the next morning to do his chores.

To the Fischer family, Mengele displayed many of the endearing qualities of the Beppo of old. He kept the family entertained with his light banter. The household took a deep liking to their new hired hand, especially the Fischers' small children. On Christmas Eve, he played Saint Nicholas for them. According to the Fischer family, Mengele gave a hilarious performance impersonating the crotchety, benevolent German Father Christmas. Mengele's reputation as a thespian was assured, and thereafter he was frequently called upon to entertain.

The only member of the household who distrusted Mengele was

the farmer's brother, Alois. Noting Mengele's odd habit of washing his hands after finishing every chore, Alois concluded he had never been a farmhand or a soldier, and figured out that their stylish employee was probably a wanted high-ranking Nazi, using his brother's house to hide out. But he didn't give Mengele away. The war criminal was able to lead a relatively peaceful life for over three years. Of course, toiling as a farmhand was a blow to Mengele's self-esteem, but his notebooks reveal he consoled himself by dreaming of his future, when he would return to his science and his experiments. Mengele's greatest source of anxiety came from not having news of his family. He didn't dare show himself in Günzburg again, but like all the refugees of this long and devastating war, he longed to know how his parents and brothers, his wife and his son, were doing.

MENASHE LORINCZI:

Neither Lea nor I knew after the war that our father was alive. We assumed he had suffered the same fate as our mother. After several months under the Russians' care at the Auschwitz clinic, Lea and I left the concentration camp and tried to find our way home, to Cluj.

There were still thousands of survivors wandering around Eastern Europe, trying to find their way back home. It was very difficult. Much of Eastern Europe's rail system had been heavily bombed. Even months after Liberation, it was not functioning properly. We would board a train, and a few kilometers along the way we would get to a river and the train would stop, unable to continue because a bridge had been bombed. All the passengers were forced to get out, wade through the river, walk to a village, and wait for days until another train arrived.

It took several months for us to make it back home to Cluj. We found one uncle—our only relative to have survived.

Unbeknownst to Lea and me, our father had spent the war in the relative safety of a Russian labor camp. He'd spent the last months of the War and Liberation mourning his entire family, certain all of us had perished at the hands of the Nazis.

Ironically, he found out we were still alive because of a newspaper article—one that quoted me at the time of Liberation. The interviews I gave to the press corps who came to inspect Auschwitz after Liberation were published all over the world. An account of my experiences under Dr. Mengele even found its way to a Russian newspaper widely read in my father's labor camp.

My sister and I were identified by name—as was our father. As a result of this story, he learned my twin sister and I were still alive.

❧

LEA LORINCZI:

Our father began making intensive efforts to locate us. He had no idea where we or our mother might be. At last, he found out we were back in our old hometown in Cluj. After he pleaded with the Russians, they agreed to release him from labor camp. He was put on a transport with sick people and began the long journey to Cluj.

One day, he turned up at our door.

The reunion was heartbreaking. He didn't know what had happened to our mother. "Where is Mommy?" he kept asking us.

❧

MENASHE LORINCZI:

I will never forget that day as long as I live. We were laughing and crying at the same time.

Father took my sister and me in his arms. Then he sat down. He put me on one knee and Lea on the other. He wouldn't stop kissing us. And we couldn't stop crying.

We didn't have the heart to tell him what we suspected: that she'd perished in the gas chambers of Auschwitz.

It took a long time before we could talk to our father about the last time we saw our mother. In the meantime, he moved in with us and we started life anew. We were so happy to be with him—but we were also so sad, because our mother was not with us.

As refugees kept coming back, we would ask them about our mother—what had happened to her, did they ever see her? Her fate was a complete mystery. The answer was always, "We don't know."

Anxious for news about his wife and family, Mengele asked his friend the doctor to visit Günzburg. Ulmann's brother-in-law was going on vacation and would be passing the Bavarian town, so he was glad to oblige. Once in Günzburg, he discreetly sought meetings with various members of Josef's family. He let everyone know their beloved Beppo was safe.

Meanwhile, the once-proud Mengele family had problems of its own. The indomitable Walburga was deathly ill; she would expire shortly after the New Year, in 1946. Mengele's father was being scrutinized by U.S. Occupation authorities for his Nazi involvement during the war. It would not be long before he would be placed in a detention camp. Lolo, Josef's younger brother, who had fought in France, was also a POW, languishing in a camp in Yugoslavia. As for Irene and little Rolf, they had moved after the war from Freiburg to the Günzburg area and were being taken care of by the Mengele family. Because of the famine that was ravaging Germany, Irene had had no choice but to leave her hometown: at least in Günzburg, with her wealthy in-laws, she could be assured of food for herself and little Rolf. But she had left Freiburg, a lively center of culture, somewhat reluctantly for stifling Günzburg. Irene's parents accompanied her and the child, and together, they rented a house in nearby Autenried.

Irene learned that her husband was surviving, but she knew little of how he was faring in these dangerous times. American Army investigators had been to Günzburg inquiring after her husband's whereabouts. By late 1945, the search for Nazi war criminals, rather than dying down, had intensified, as the Allies made preparations for war-crimes trials. Specialists at the U.S. Army Counterintelligence Corps, whose regional headquarters were in Augsburg, thirty minutes away from Günzburg, joined the search, as did many other units and agencies.

As survivors emerged to talk about Auschwitz, Dr. Mengele's name was heard more and more frequently. Although many doctors had served on the selection ramp of Birkenau, victims seemed to recall Josef Mengele most vividly, hence the start of the legend surrounding the Angel of Death. The image of the elegant young SS doctor who whistled and smiled as he sent people to die was indelibly imprinted in their memories.

Only the twins remained silent. Several had made it back to their native villages and moved in with distant relatives. After their ordeal with the Nazis, they were relieved to be home, although their homes had changed: They were now under the control of the Communists. Some were scattered in relocation centers and orphanages throughout Europe, awaiting transport to Palestine. Still others were in Catholic convents being cared for by nuns, along with other war orphans.

But all of Mengele's twins were in mourning. They grieved not only for their loved ones, but also for their lost childhood, which had been so cruelly snatched away, and would never be recaptured.

ZVI THE SAILOR:

As children before the war, my brother and I fought all the time. Our mother would tell us, "You will cry one day, and it will be too late."

But I never cried. I know that after the years I spent in the camps, the refugee camps as well as the concentration camps, I could not be a child. Because if ever once I had acted like a child, I would have died.

After Liberation, many different refugee organizations asked me where I wanted to go, what I wanted to do. I decided I had to try to go back to my native village. I hoped I would find my parents, or someone—anyone—I had known.

After about six or seven months, I finally made it back to my hometown, and I met up with my brother. We quickly learned that out of hundreds, thousands, of Jews who had lived there before the war, only one or two had returned.

And absolutely no one from our own family had come back—neither our parents, nor any of our six brothers and sisters. We were the only survivors.

My twin brother and I stood in front of our house and we cried—we cried and cried.

And then I remembered my mother's warning, so many years before—"You will cry, and it will be too late."

5

THE TRIAL THAT NEVER WAS

Judith Yagudah:

They were awful, those years between 1945 and 1950.

Mother and I settled in our old hometown. Mother was even able to find some of her old friends, from before the war. All were survivors, and they constantly talked about Auschwitz.

They would talk about the members of the family who had returned, and those who hadn't. Each family had its wounds. Each family counted its dead. It was awful for me to listen to them.

I was a young girl, but I was much more serious, much more moody, than other girls my own age. I was not like the others. I had trouble concentrating in school. It was very hard for me to study.

In the beginning, I thought about my sister, Ruthie, a lot. But then, less and less. Life has to go on, I guess.

But Mother took the loss of Ruthie very badly. She had been a doting mother of twins. She came back with only one child: It broke her heart. For one year, she did not go out anywhere. Mother was a young woman after the war, but she never remarried.

Irene Shoenbein Mengele cut a striking figure as she walked the streets of Günzburg, tall and sorrowful, her blond hair glistening against her black mourning clothes. Contemporaries recall how well she took to the role of war widow, how convincingly mournful she looked.

Not yet thirty, Mengele's wife was still young and attractive, with that tall, blond, buxom beauty that had made her such a desirable mate to her race-conscious husband. As she walked, holding little Rolf by the hand, it was said in the Bavarian town that she caught the eye of several of the young men. But Autenried and Günzburg held no charm for her, and even though she was well-provided-for by the Mengeles, she was anxious to return to her beloved Freiburg, famine or no famine.

Although the Mengele family knew by the late summer of 1945 that Josef was alive, they had decided to pretend he was missing and apparently dead. This was clearly the easiest way to fend off troublesome inquiries. The ongoing searches for war criminals and the proximity of the U.S. Occupation forces was a constant source of anxiety. The Mengeles could only hope that with tens of thousands of Nazis on the wanted lists—many of them far more prominent than their Beppo—investigators would be only too glad to strike off another name as missing or dead.

For the Mengeles, protecting Josef became both a game and an obsession. By a strange twist of fate, American forces decided sometime in 1945 to move into the Mengele mansion. The family rushed over to the villa, and while pretending to collect family valuables, they scooped up any photographs of Josef that were lying around the house. There were quite a few; the brilliant young doctor was the pride of the Mengeles.

An investigation by the Justice Department some forty years later would reveal that the American occupiers developed a pleasant rapport with the caretakers of the mansion and the Mengele clan in Günzburg. They apparently never thought to interrogate the Mengeles or search for clues that could lead them to the scientist-torturer of Auschwitz. The Americans stationed in Günzburg "may not have realized who Mengele was, and the magnitude of his war crimes," according to historian David Marwell, who conducted the Justice Department probe. One day, a fire broke out at the Mengele factory. The U.S. commander personally rushed over to help put it out.

In fairness to the American forces, the man who would later emerge as the world's most wanted Nazi killer was considered a small fish immediately after the war. The British, Americans, Poles, and Russians who were conducting the manhunts wanted to punish the individuals whom they felt were responsible for the war machine. Many of the most prominent of Hitler's henchmen—such as SS and Gestapo chief Heinrich Himmler, his second in command, Reinhard Heydrich, and Propaganda Minister Josef Goebbels—were already dead. But even compared to the surviving architects of Germany's reign of terror, like the minister of war production, Albert Speer, or Foreign Minister Joachim von Ribbentrop, the Auschwitz death-camp doctor seemed a minor player.

The officials who were putting together the first major war-crimes trials decided to start by prosecuting the Nazis who had ordered the death camps to be built, not merely those who had worked in them. The trials were to be held in November 1945 in Nuremberg, the birthplace of the original 1933 racial laws Mengele had so admired. The Office of the U.S. Chief Counsel for War Crimes, in charge of overseeing the trials, sent out its own investigative teams to scour the German countryside for wanted Nazis. Because of the need to begin as quickly as possible and the vast number of war criminals still at large, there was a hurried, haphazard nature to the manhunts. The teams were also hampered by the utter chaos of postwar Germany: even back then, it was common knowledge that major war criminals were slipping out of the country.

As Benjamin Ferencz, then a twenty-five-year-old attorney running the Berlin Office of the U.S. Chief Counsel for War Crimes, later recalled, the emphasis was on speed. Once a Nazi was caught, a case against him was hastily prepared, paving the way for a trial.

The opening of the International Military Tribunals sparked worldwide interest. In the course of the next two years, twelve Nazis who had been far more powerful than Josef Mengele were sentenced to die for their crimes against humanity. Three others received life sentences, four were given jail terms of ten to twenty years, and three were acquitted. One of the most notorious of Hitler's cronies, Martin Bormann, was tried and sentenced to death in absentia. Even though the trials had been put together in a slapdash way, by young and often inexperienced prosecutors, they ultimately did achieve their desired result—the punishment of the leaders of the Third Reich. Some of those men stiffly took the stand and argued they had only "followed orders." Others, like Hermann Göring, committed suicide rather than face the public reckoning.

MOSHE OFFER:

I didn't want to live after I came out of the camps. I couldn't bear the notion that my entire family was gone.

After Liberation, I developed all sorts of nervous ailments. I began having fits. I had trouble sleeping. I couldn't concentrate. The doctors said it was the result of all the experiments I had undergone at Mengele's hands.

I also began to have problems with my memory. There was a lot that I couldn't remember—a lot that I did not want to remember, I guess.

I was completely broken down, completely sick. I felt I did not have the courage to continue living. I was always thinking about Tibi, about how he had died, about the horrible tests and surgeries Mengele had performed on him. I could not get over what had happened to Tibi, how they had taken him away from me. I could not get over the death of my twin brother.

I could not get out of my mind all that Mengele had done.

Mengele, safely ensconced in his Mangolding hideout, viewed the Nuremberg proceedings with unvarnished contempt. In his autobiographical novel, first begun as his memoirs and later "fictionalized" by his changing of the characters' names, the protagonist likens the trials to "political theater." Nuremberg, says Andreas—Mengele's persona—is nothing but a "farce," a feeding frenzy for a world exultant at seeing Germany brought to its knees. The former death-camp doctor apparently saw no value in punishing the worst offenders of the Hitler regime, for in his opinion they had done nothing wrong except to lose the war. During one of his many monologues about the proceedings, Andreas excuses Hitler's actions, saying he "only did what others before him had done."

Mengele's name was mentioned a number of times during the Nuremberg deliberations. The former commandant of Auschwitz, Rudolph Hess, briefly discussed Mengele's death-camp career, stating that he had conducted experiments on twins. A former Auschwitz inmate, Marie-Vaillant Claude Couturier, described Mengele's cheerful, effortless way of performing selections, his habit of whistling classical tunes, and, of course, his obsessive interest in twins.

A perusal of Mengele's writings (in addition to his novel, he left behind diaries and notebooks) indicates that Mengele remained unrepentant, a steadfast Nazi. To the end, he remained convinced of the necessity and importance of Germany's actions. Bitter and angry over his country's defeat, he even mused about ways to start a resistance movement to oust the American Occupation forces. But he knew it was a lost cause. The German people might have felt loyal to the Third Reich, but they were not about to risk more death and devastation to bring it back.

Nor was Mengele disturbed by the scope of the crimes against the Jews that was revealed during the trials. To the contrary, he resented the world's uproar over details of the Final Solution, and justified the existence of the death camps. Mengele felt the Jews were to blame for the war, and hence for the cruel fate they later encountered. In his novel, his hero says the Jews had pushed the Allies into fighting Germany by spreading false propaganda about the Nazis. "Hitler warned Jews in his speech in 1939 not to stir up the people with war against Germany, that it could end up badly," his protagonist, Andreas, tells a friend. Spouting classic anti-Semitic dogma, he decries the "internationality" of the Jews and their supposed links with the intelligence services of other countries. "This fact alone would be enough to take measures such as the concentration camps." As "potential enemies of the state," the Jews had to be killed.

PETER SOMOGYI:

The Communist regime in Hungary became increasingly hostile to the Jews: It considered us "class enemies." The government knew anti-Semitism was on the rise, but nothing was done to stop it. The new rulers simply did not like the Jews.

In the first months after the war, my twin brother and I assumed our father had died. We had no idea what had become of him: that he had been interned at Dachau.

Then, in July, we got the news that he was alive. After Dachau, he hadn't wanted to return to Hungary. He was certain that his entire family was dead. But someone told him that his sons were alive, and so he returned to our hometown.

Father showed up in Pécs in August 1945. We hugged and kissed. He asked us many questions, but we were very reluctant to answer him. We didn't want to tell him what had happened to our mother and sister.

We started our life yet again. The three of us moved back to our old house. We cleaned the place up and bought new furniture. My father opened a store selling automobile parts—exactly like before the war. In other words, we tried to pick up the pieces.

But our lives were marred by the political upheaval around us. Hungary was very turbulent. There was one regime after another. Then the Communists took over. It was stable—but extremely difficult for the Jews.

Father, who was very clever, was quickly able to make a success of his business. But then a Communist group inside our school decided that we were too rich. It was the old brand of anti-Semitism. Once again, the Jews had too much money. We were charged higher tuition than the other students.

The other kids ganged up on my brother and me. They called us "capitalists"—the new insult. The children had inherited the anti-Jewish feelings from their parents. They had been raised to believe that Jews were evil.

I was dragged into school court. I was told that I was not a true member of the proletariat. I was accused of being a capitalist landowner—even though my father owned no land.

Then, we were even subjected to the old, prewar insult: "Dirty Jew."

❧

EVA MOZES:

After we realized no one in our family had survived, Miriam and I went to live with an aunt in Cluj.

This aunt had lost both her husband and her twenty-six-year-old son in the war. She was always in mourning for her son. We could not possibly replace him. She never gave us a hug, or kissed us.

Miriam and I busied ourselves with schoolwork and tried to forget the war.

When the Communists were taking hold of Romania, we became very involved with the Communist party. We joined the Young Pioneers in my school, and because Miriam and I got along very well with the

Communists, they put us in charge of our school's Communist party. We were called "Pioneer leaders."

The Communists made us feel very special. They spoke about freedom and liberty. They would talk about the horrible Germans, the crimes of the Fascists. It all sounded wonderful. I felt very good about Communism.

But then some rumors began circulating around Cluj that a Jewish vampire was sucking the blood of Christian children.

After that, life in Romania became much more difficult.

❧

LEA LORINCZI:

We began to suffer under the Communists.

We were very religious, and the Communists didn't want religion.

My twin and I attended a Jewish school, just as we had before the war. We worked very hard.

But the Communists closed down our Jewish school. They forced us to attend classes on the Sabbath.

It was awful. It brought back the terrible years before the war, when there had been so much anti-Semitism.

JUDITH YAGUDAH:

I felt the anti-Semitism at school very keenly. At first, I attended a Jewish school, and was protected. Then, the Communists engineered what they called an "educational reform." They shut down the Jewish schools.

In my new school, nobody wanted to be my friend. The Christian boys and girls would have nothing to do with me.

I felt all alone.

As the International Military Tribunal at Nuremberg continued its grim business, and as death-camp survivors emerged from the shadows to testify, Mengele's name began to appear on the wanted lists of more and more countries. Perhaps the Nuremberg Tribunal, concerned as it was with statesmen and generals, would not have been the proper forum to try Mengele even had he been captured. But Mengele would have made an ideal defendant for the British-run Belsen trials. Held in Lüneberg, site of the old Bergen-Belsen concentration camp, these trials took place at the end of 1945, at the same time as Nuremberg was getting under way.

Many of the Lüneberg defendants had served both at Auschwitz and at Bergen-Belsen; among them were many of Mengele's old cronies, including some SS doctors. One of them, Dr. Josef Kramer, had been a high-ranking physician at Auschwitz-Birkenau. According to former inmates, the imperious, thuggish Kramer was one of the few men at Auschwitz—perhaps the only man—that Dr. Mengele feared. However, the "star" defendant of the Belsen trials was not Kramer but Irma Grese, Mengele's old sidekick.

After the fall of Auschwitz, Irma had sought to continue her death-camp career at Bergen-Belsen. There, too, she had quickly distinguished herself for her wanton cruelty. She would beat prisoners with her own hands, sometimes until they fell unconscious and died. Only at the end, amid rumors that British troops were about to overrun the camp, did Grese make an about-face and treat prisoners with a semblance of kindness. She even dared ask them to put in a good word for her. But her entreaties did not help. When the British entered Bergen-Belsen and saw the mounds of dead and dying, Grese was rounded up with other Nazis.

During the trial, prosecutors wondered how such "a very young girl" could have committed acts of such "very great savagery and cruelty." When Irma took the stand, she seemed curiously subdued. She readily admitted to many of the accusations leveled against her. In her own defense, she only pointed out that she had acted with the approval of higher authorities—most notably, Dr. Josef Mengele. She had "always" accompanied Mengele; he had been her boss.

Mengele's fair companion was hanged, along with Dr. Kramer.

Lüneberg was the first, but by no means the last, of the many war-crimes trials from which Mengele would be conspicuously absent. The proceedings did establish early on a record of his crimes at Auschwitz. By the end of the Lüneberg trial, much was already known about his actions as an SS camp doctor. Even amid the panoply of Nazi horrors, Mengele stood out for the sheer magnitude of his crimes.

TWINS' FATHER:

I was shocked when I was reunited with my sister, Magda. We met after I had left Munkács and settled in a different city, in Romania.

I worked as an accountant, and someone told my brother-in-law where I was.

My twin sister had been such a beautiful girl before the war. But she was completely changed, and in a terrible way. She had a beard all over her face—a real man's beard.

It was horrifying to look at her, and to remember how she had once been, so radiant and happy with her young son—the little boy Mengele sent to the gas chambers.

❧

MAGDA SPIEGEL:

I was very broken down after the war, very sick.

I was able to locate my husband, and we were quickly reunited. He had spent the war in labor camps all over Russia. We decided to settle in Czechoslovakia.

I missed my son. I wanted a child very much. But I did not have my periods. I was very frightened. I didn't know what they had done to me at Auschwitz—nobody knew.

I went to many doctors. They told me, "Wait."

After a long time, I started functioning normally again. I became pregnant, and gave birth to another boy.

But I continued to think about Auschwitz. I was always having nightmares about the camp.

The Nuremberg and Belsen trials only intensified the Allied powers' desire to punish the criminals of the Third Reich. As new, exhaustive manhunts got under way, Mengele was forced to remain in hiding under his alias of "Fritz Hollman." Despite the security Mangolding offered, Mengele was unhappy at the Fischer farm, where the work was tedious and exhausting. He could not get used to the harsh regime of physical labor and developed severe shoulder pains. His main distraction continued to come from books. And whenever he could, he visited his friend, Dr. Ulmann's brother-in-law, where the two discoursed on everything from medicine to the affairs of the day—specifically, the Nuremberg trials—and even talked about Mengele's future hopes. Unbelievably, he somehow still thought he could resume a normal life and become a university professor, like his old mentor Verschuer.

Mengele's cunning old patron was faring considerably better than his former assistant. Although he had been the force and the inspiration behind Mengele's deadliest experiments at Auschwitz, Verschuer never had to hide out on an isolated farm. On the contrary, he was highly visible as he sought to get his old job back as head of the reconstituted Kaiser-Wilhelm Institute being formed at the University of Frankfurt.

Verschuer was as much an opportunist after the war as he had been under Nazi rule. In February 1945, as the Russians were approaching Berlin, he had fled to his hometown of Solz, taking the institute's entire library with him. This included eighty-eight cases of books on heredity and racial science, along with fifteen cases of books from his personal collection, most of which he hid in small inns around Solz.

It is also believed that Verschuer hid or destroyed any evidence that could be used against him, including his voluminous correspondence with Mengele at Auschwitz. As news surfaced of what his protégé had done at the death camp, Verschuer feigned surprise and indignation. He claimed to the press that he had no knowledge of selections and gassings, grotesque experiments and mass murders. Certainly, he had no idea how Mengele had gathered the grisly twins' specimens that had been sent to him personally at the Berlin Institute. Verschuer said he had assumed that the slides and tissue samples, organs and skeletal remains, had come from people who had died "of natural causes."

In order to make sure he got his old job back, Verschuer made Mengele the scapegoat, passing off responsibility for any scientific excesses onto his protégé; just how Mengele felt at seeing himself blamed by his former idol is unknown: There is no apparent reference to Verschuer in any of his papers. But Verschuer also resorted to many other lies to regain entry into the academic fold. For example, he told the mayor of Frankfurt, who was overseeing the opening of the university, that he was close to developing a vaccine for tuberculosis. It was an ingenious idea: In the postwar months, an epidemic was raging throughout Germany, and there was an all-out search for a vaccine or a cure.

Forced like thousands of other Germans to undergo "Denazification," whereby former Nazis were probed for the extent of their involvement with the Reich, Verschuer was able to con the Americans who presided over the procedure into thinking he had been little more than a victim. If the Denazification authorities had found him to be an active Nazi, criminal charges would have been filed; as it was, Mengele's mentor was declared to have been only a "hanger-on" of the Nazi regime, and fined the trifling sum of six hundred marks. His appeal was also helped by the accolades of his former colleagues at the University of Frankfurt, who heartily endorsed his efforts to regain his former chair.

Like his favorite student, Otmar von Verschuer had no regrets about his commitment to eugenics, or about what he had done in the name of science. In years past, he had raised the pitch of his scientific preachings to appeal to the Nazi rulers. Now, Verschuer knew exactly how to soft-pedal his views to gain the trust of the postwar establishment.

By the end of 1945, while the fugitive Mengele was sorting potatoes and cleaning stables, his teacher was on his way to attaining his former prominence. Verschuer was able to slip, chameleonlike, into the protective coloration of an innocent academician who has been trapped in evil surroundings. While men who had been far less influential in the Nazi regime than he now feared for their lives, Otmar von Verschuer worried only about restoring his lost reputation.

Verschuer's self-serving technique was useful to other ex-Nazis as well. Mengele rapidly became the scapegoat not only for his old professor, but also for many of the theorists of racial science. They pretended the Auschwitz doctor was an aberration who had distorted the "ideals" of eugenics. Despite the fact that their theories had been used to justify the killing of millions of "inferior" human beings, they chose not to condemn these theories, but only the man who had best put them into practice. They rejoiced when Mengele was thrust into the limelight, for it meant their own actions would remain obscured.

Verschuer's attempts to return to the University of Frankfurt, which were going so smoothly in 1945, ran into trouble in 1946. Two of his more ethical former colleagues, upon hearing of his possible reappointment, published a lengthy article on May 3, 1946, in the *Neue Zeitung*, a Frankfurt daily, describing Verschuer's activities during the war. The two scientists, who had worked with Verschuer for many years, exposed the fact that he had corresponded with Mengele and knew exactly what was going on at Auschwitz. They charged that he had known about the layout, functions, and activities of the death camp, and was well aware of the source of the "specimens" he regularly received from Auschwitz: Jews put to death by Mengele. As an intimate of the Nazi hierarchy in Berlin, Verschuer knew about the Final Solution, had advocated killing Jews as a eugenic measure, and, the doctors claimed, also knew precisely how the exterminations were being carried out, since they believed he himself had visited Auschwitz.

The scandal that resulted from the publication of the *Neue Zeitung* article aborted Verschuer's plans to resume his career in Frankfurt. More significantly, the ensuing brouhaha brought both Verschuer and Mengele to the attention of U.S. authorities at Nuremberg. As the international military tribunals were winding down, plans were being made to conduct additional trials. There was to be a special proceeding for individuals who had worked in the death camps.

At the Berlin office of the U.S. Counsel for War Crimes, a young investigator named Manfred Wolfson was assigned to look into the bizarre affair of Doctors Mengele and Verschuer. Despite the large backlog of cases, Wolfson began digging into the two men's backgrounds. He interviewed their former colleagues from Kaiser-Wilhelm and quickly grasped that Verschuer was as important a target as the more notorious Auschwitz doctor. As Wolfson probed Verschuer's past, he discovered the professor had consistently used his position as a racial scientist to enhance his ties with the Nazis. He also noted the frequent anti-Semitic, pro-Hitler references in articles Verschuer had published in the thirties and early forties.

In a 1946 report Wolfson carefully prepared for his superiors, he made it clear that most of the horrors for which Mengele was becoming famous had been perpetrated under the aegis of Verschuer. Wolfson wrote that even Mengele's passion for experimenting on human eyes could be traced back to Verschuer. His old professor had been studying the development of pigmentation in eyes, and had a keen interest in different-colored or "heterochromatic" eyes. To satisfy Verschuer's needs, Mengele would gas inmates who possessed this unusual trait, then ship the eyes to him in Berlin.

These gruesome discoveries prompted Wolfson to recommend that Verschuer be "interrogated and tried." Only after exposing Mengele's mentor did Wolfson pursue his former protégé. He cited witnesses who described Mengele's obsessive interest in pseudomedical tests on inmates. He noted the camp doctor's fascination with the sterilization of women. And, of course, he described Mengele's passion for young twins. "Twins and triplets, predominantly children, were kept in separate barracks so they could be experimented on properly," one survivor had told Wolfson. This man recounted how his own twins had been placed in Mengele's hands upon their arrival in Auschwitz in July 1944. He had asked Mengele whether he would ever see his children again. "Of course!" Mengele reportedly replied, cheerful as ever. But as the grieving father mournfully observed, "To date, I have not seen anyone, nor had any news."

MENASHE LORINCZI:

For one year after the war, we had no news about our mother. As refugees kept coming back, we would ask them if they had seen her: Was she here? Was she there? Finally, by talking to dozens of people, we were able to piece together what had become of her.

Mother had been part of a group of traveling women workers. She had spent some time working at Auschwitz, then had been transferred to several other labor camps around Poland.

At last, she ended up in Riga, a death camp in Latvia. What the Nazis couldn't do at Auschwitz, they succeeded in doing in Riga: They simply killed everybody.

They put all the women and children on a ship in the middle of the Baltic Sea, and sank it. Everyone drowned.

My sister couldn't accept the fact that Mother had died. Lea kept believing Mother had somehow survived. We both did.

We would think, "Maybe she is still in Russia."

❧

LEA LORINCZI:

I cried myself to sleep every night those first years after the war, thinking about my mother. I often dreamt that Mother had come home.

Then, I would get up and realize it was only a dream, and I would start crying again.

It was a period when I felt very sad. I did not want to be with other people. My father was also depressed. He missed our mother very much, and always wanted to talk about her.

❧

MENASHE LORINCZI:

My sister was very unhappy, and she tried to forget her sorrow by burying herself in the Communist movement. She was extremely active in it. She was not a very strong-minded person—she was easily suggestible, and the Communists brainwashed her.

Lea wanted me to join, too, but I knew the Communists were no good. I knew they hated the Jews.

They always claimed to attack only "Zionists." But who were the "Zionists"? The Jews. Only the Jews.

I finally had to forcibly yank my sister out of the movement.

After that, I knew that we had to leave Hungary. We had to emigrate to Palestine.

But the Communists wouldn't let us go.

At the conclusion of his report, which Wolfson submitted to his superiors in November 1946, the investigator recommended that "SS Haubsturmführer Dr. Josef Mengele be placed on the wanted list and that he be indicted for war crimes." At last, the Angel of Death seemed about to get his due.

But Mengele, tucked away in his pastoral hideout, could not have been more removed from the machinations to capture and try him. He did not know of Wolfson's report, and was leading a quiet, peaceful life. His major source of anguish stemmed from his longing to be reunited with his family.

Mengele's homesickness was relieved by a visit from his brother Karl in October 1946. The reunion was bittersweet. Karl confirmed the sad news of their mother's death and their father's imprisonment. Karl brought with him Hans Sedlmeier, an old school chum of Josef's, who was now managing the farm-equipment factory. The Mengeles had been forced to turn over control of the company to Sedlmeier because of the Americans' interest in disbanding any firms run by ex-Nazis. Sedlmeier was an ideal interim manager, both because of his loyalty to the Mengeles and the fact that he did not have a Nazi past.

The visit of Karl and Sedlmeier broke Mengele's isolation. The two men brought Mengele up to date on Günzburg gossip as well as on how the family business was faring. And afterward, there was a steady stream of visitors to the Mangolding farm area. One day, Irene herself showed up. Their reunion was a curious mixture of tenderness and disaffection. Years later, Mengele relived the meeting in his autobiographical novel: "I never believed that I would see you again," Irmgard, the character who represents Irene, tells Andreas. She had apparently lost all hope of being reunited with her husband.

Irene advised Mengele to leave Germany at once; it was simply too dangerous to remain there, she believed. She told him how a few months earlier an American officer and his interpreter had turned up at her house, demanding to know where Mengele was. Although very nervous inside, she had managed to retain her composure. Flashing her nicest smile, Irene had told them the stock story: that Josef Mengele was "missing" in the Eastern Campaign, and was presumed dead.

While the officer had listened sympathetically to the pretty young woman, standing there with her young son, the interpreter—a Jewish soldier—had been more skeptical. He had lashed out at her, informing her that Mengele was responsible for the deaths of millions of Jews. The two men had finally driven off, but it had taken all of her energy to convince them she really knew nothing of her husband's whereabouts.

The tension of having to cover for her war-criminal husband affected Irene deeply, and added to her general unhappiness with her marital situation. Since their marriage, weeks before the start of the war, Irene and her husband had spent little time together. She had stoically borne Josef's going off to war and, after that, to Auschwitz for his scientific career. But now, peace was here at last, the country returning to normal life—yet still they were apart. Her husband was a fugitive, wanted by several governments for his alleged wartime activities. As their son, Rolf, would later reveal, although Irene didn't believe the accusations leveled against Josef, she found it frustrating not to be with the man she still loved, to be condemned to live by herself, with none of the joys of family life she craved.

TWINS' FATHER:

After the war, all the survivors wanted to get married, to build homes, to forget the Holocaust and what had happened.

I met my wife seven months after Liberation. She came from a small village in Czechoslovakia. She had also been interned in the camps.

We courted exactly four weeks.

We were married on January 27, 1946—exactly one year after Liberation. My first child was born less than a year later.

We lived in a beautiful city. I had a good job and a very lovely house. Yet neither of us felt settled.

❧

VERA GROSSMAN:

When Mother came back to her hometown after the war, she learned that no one in her family had survived. Her parents and her seven brothers and sisters had all gone to the gas chambers. She was completely alone, with two seven-year-old daughters to care for.

The three of us went back to our old estate—but the people who had taken it over threatened to kill us. And so we fled to another town.

Mother got together with another Jewish woman, also a camp survivor. They rented two small rooms. They would take in geese and feed them—fatten them up—then sell them for a higher price. People loved buying fat geese, because they could get schmaltz *—a thick, delicious paste you make from the fat.*

Eventually, Mother was able to buy a cow with the money she earned fattening the geese. She was convinced that as long as she had milk for her twins, we would never be hungry again.

But she was always very nervous. She worried about who would take care of us if something happened to her.

One day, a man she had met at Auschwitz turned up at our door. After Liberation, he had gone off to search for his family. He traveled throughout Europe, to Hungary, to Poland, to Czechoslovakia, in search of survivors. Only after he had determined that no one had survived did he come to us and ask for Mother's hand.

They were married very quickly. Mother said yes because she felt he would be good to Olga and me, and take care of us. And he did. He missed his own family, and was very loving toward us.

We moved to another town. There, we rented a large apartment. Mother stopped working and had a baby. Then she had another baby very quickly afterward. We were a family again.

But there were a lot of rats in that new apartment. It reminded me of Auschwitz. Each time I saw a rat, I thought of Auschwitz.

❧

TWINS' FATHER:

I found I couldn't feel safe anywhere in Eastern Europe—not after what had happened.

I simply didn't trust the land where I was living. I had a feeling there was nowhere in Eastern Europe I could really settle down and establish roots.

My wife and I both wanted a place we could call home. We wanted to emigrate to Palestine.

All we could think of was running away from Europe and all its memories.

We wanted out.

To Mengele's shock and dismay, during her visit Irene told him she wanted out of the marriage, an idea she had obviously been contemplating for some time. A divorce would free her from the innumerable burdens—and few rewards—of being Dr. Mengele's wife.

Mengele was distraught. "She wants to leave me because I am not back home like the other husbands—as if it were my fault," the protagonist in his book complains at one point. Mengele was unsympathetic to his wife's grumblings. The doctor-turned-farmhand had troubles enough of his own. In his book, there is a scene where Andreas berates his wife, sternly reminding her of the thousands of women whose husbands are POWs and who are in equally dire straits.

The reunion with Irene left Mengele feeling depressed and forlorn. In the manuscript he wrote decades later, he described his sadness at losing both his mother and his wife. Walburga's death clearly caused him the most pain, however. "One can never replace a mother," he observes at one point. But the prospect of losing Irene was almost as distressing. It upset Mengele's dream of once again leading a traditional family life, even though the life he could now offer her hardly fit the typical bourgeois mode.

Although Irene's advice to leave Germany made eminent sense in this period of manhunts and war-crime trials, Mengele felt safe enough in his farm retreat to risk staying put. Did he really think his notoriety would fade? Could he possibly have believed that the Allies would leave Germany and forget all about him? The smug former dandy of Günzburg was certainly deluded enough to have entertained such fantasies. Perhaps in his heart he even hoped he would be able to reemerge and resume his old life in his native land. But for now, he was also pragmatic enough to take no chances and remain in hiding.

HEDVAH AND LEAH STERN:

Life in our hometown in Hungary was very disappointing, very sad. We cried all the time.

Only two widowed uncles had survived from our entire family. We were taken to live with them.

These uncles were both very religious and very strict. We quickly realized there was no future for us in Hungary, there were no young people left. There was no Jewish community.

There was nothing—absolutely nothing.

We dreamt of going to Palestine, but our uncles forbid us to join the local Zionist movement, the B'nai Akivah. They didn't want us in any coeducational groups. We finally were able to find another Zionist group that admitted only girls: Our uncles allowed us to join that.

We decided we wanted to build a new future for ourselves—in Palestine.

The continuing Nuremberg proceedings, in particular the concentration-camp doctors' trial, should have made life most precarious for Mengele. In the fall of 1946, twenty-three representatives of the Nazi medical establishment were indicted. Of these, twenty were physicians, while the rest had served Nazi science in administrative capacities. It is ironic that this trial, where Mengele ought to have been the star defendant, came and went without him; the Wolfson memorandum, submitted even before the trial opened, had carefully outlined the medical crimes both of Mengele and Verschuer, urging that they be indicted.

Somewhere along the way, however, the report disappeared. What became of it is not known, even to Wolfson, who is now teaching languages for the U.S. Army, and has settled on the West Coast. It may have simply fallen through the cracks, although one can conjure other, far more sinister explanations.

To this day, that Mengele escaped judgment—even in absentia —in the trial of doctors like himself remains a great mystery of the postwar era. More than any of his peers, he exemplified the excesses of Nazi medical science. He engaged in far greater atrocities than Dr. Waldemar Hoven, the man picked by the Nuremberg team as the premier example of death-camp physician. Hoven, chief doctor of Buchenwald, had participated in many selections—but he never conducted medical experiments. Conversely, those doctors on trial for their inhuman use of human "guinea pigs" had not selected victims for the gas chambers. The Angel of Death of Auschwitz was one of the few Nazi doctors who had both relished selection duty and performed human experiments.

"If they could have gotten hold of Mengele, there is no doubt he would have been tried and sentenced to death," observed the late John Mendelsohn, a leading authority on the Nuremberg Trials who worked at the National Archives in Washington until his death in 1986. Mendelsohn believed that Mengele's case was so outstanding, the sum of his crimes so horrifying, that he should have been tried along with the statesmen and generals before the international military tribunals.

At the doctors' trial, experts reached all the way back to ancient Greece to find a suitable name to describe the perversions of German science under Hitler. They called it "thanatology," the science of death, after Thanatos, the Greek spirit who personified death. The proceedings revealed that over two hundred German doctors had been direct participants in "research" crimes, while hundreds, perhaps thousands, more had stood silently by. It was clear that no useful scientific findings had resulted from these experiments—only the suffering and deaths of helpless human beings. At Buchenwald, doctors had searched for a cure for malaria by exposing inmates to the disease-carrying mosquitoes. The prisoners were then "treated" with large doses of questionable drugs, which invariably killed them. At the trial, female witnesses tearfully recalled the brutal sterilization procedures they had undergone at Auschwitz and Ravensbrück.

But oddly enough, no witness or prosecutor mentioned the supreme thanatologist, the Angel of Death, Dr. Mengele. No one mentioned the medical experiments he'd performed on twins, triplets, dwarfs, and giants. The trial never discussed his diabolical tests, castrations, and surgeries. And not a word was said about the active cooperation between Mengele and Professor Verschuer, one of the leading theoretical scientists of the Third Reich.

The answer to this great mystery of why Mengele's name never surfaced may lie in the sheer immensity of the medical crimes confronted by the Nuremberg prosecutors. According to Neal Sher, director of the U.S. Justice Department's Nazi-hunting unit, "It is a commentary on the barbarity of the Nazis that someone with as much blood on his hands as Mengele would not have been Number One on the prosecutors' lists. That fact alone should put into perspective the unbelievable scope of the horrors."

Another Holocaust expert, Dr. Robert Wolfe, director of captured

German records at the National Archives, offers a simpler explanation: "He slipped through our fingers." But in defense of the Nuremberg team, Wolfe argues that "in this chaotic situation, it's remarkable we caught as many as we did. It's remarkable we even had a doctors' trial, because people were getting tired."

Sure enough, by late 1946 and 1947, war-crimes trials were starting to come under attack in the U.S. Congress. American foreign policy was changing. The new enemies were no longer the butchers of the Hitler era but the Communists. If fighting them meant joining forces with ex-Nazis, SS men, Gestapo agents, and German industrialists, the United States was ready to do so. American funding for Nazi prosecutions was drastically cut, while funds were poured into recruiting agents to aid in the fight against Communism.

EVA MOZES:

There was a lot of anxiety in the Jewish community about what the Communists were doing. At first, compared to Auschwitz, life in Eastern Europe seemed like heaven. We didn't mind standing six hours to get a loaf of bread. Or waiting two days to buy a winter coat.

At least we were free.

But then strange things started to happen.

Our house was raided a few times. My uncle was taken away by the secret police without explanation. Other people were also picked up and disappeared.

The Romanian government, faced with massive numbers of Jews who wanted to leave, tried to deal with the situation in their own way. They banned Zionist organizations. They shut down Jewish schools.

At one point, my aunt built a false closet in the house—exactly like the ones people had during the war. It was very unpleasant, very tense.

Our aunt decided there was no future in Romania. We applied for visas for all of us to go to Palestine.

❧

MENASHE LORINCZI:

Jews simply didn't enjoy the same status as non-Jews under the Communists. We were not treated equally.

The government shut down the Jewish schools. I joined a Zionist agricultural camp in Bucharest. I was taught to be a farmer. I learned how to drive a tractor, how to plant crops. All of us at this school wanted to go to Palestine and were preparing for the day that we would live there.

But within a few months, the Communists came, boarded up the camp, and sent us all back home.

I knew then that I had to go to Palestine immediately. I desperately wanted to go, but the Communists would not let me emigrate.

In Augsburg, near Günzburg, the regional headquarters of the U.S. Army Counterintelligence Corps recruited Klaus Barbie, the infamous "Butcher of Lyons," as an American agent and operative in April 1947. As head of the Gestapo in Lyons, France, Barbie had shipped groups of Jewish children to Auschwitz—among other misdeeds. Yet, instead of being tried for war crimes, Barbie was placed on the U.S. government payroll and treated like a prize source of information on postwar Communist infiltration of Germany.

To those who question whether the Auschwitz doctor also benefited from the same special treatment, it can be asserted that Mengele had little, if any, intelligence value for American agents. Genetics was simply not an area of compelling interest to them. It is probable, however, that Verschuer, deft politician that he was, succeeded in convincing U.S. officials that Mengele was a useful contact. Years later, Manfred Wolfson could only conclude that Verschuer had somehow wielded his influence to have the charges, as enumerated in the investigator's pretrial report, dropped. Wolfson learned that Verschuer had become friendly with key Americans running the Occupation government and reportedly kept suggesting to them that the colleagues who were discrediting him were "Communist agents."

By the conclusion of Poland's second set of Auschwitz trials in December 1947, the possibility of new war-crimes proceedings seemed remote. America and postwar Europe were anxious to bury the past and get on with rebuilding.

But there were some people who refused to forget what the Angel of Death had done. In 1947, Gisella Perl, the Jewish doctor who had been Mengele's assistant, published her shocking account of life at Auschwitz. Perl, who had made her way to the United States after the war and resumed practicing medicine, was haunted by her memories of the death-camp doctor. Her book focused on Mengele, and she told story after story of his atrocities at the concentration camp. In graphic, vivid prose, Perl wrote how Mengele had destroyed the lovely young Jewish girl Ibi Hillman, simply because she was so lovely. She movingly described the women of Birkenau's desperate ploy to please Mengele, and survive his selections, by smashing bricks and applying the red powder as a "rouge" to bring color to their pale faces. And she told of his unrelenting brutality toward anyone who stood in his way.

Perl was under the impression that Mengele was to be tried at Nuremberg—certainly not a farfetched assumption in the wake of front-page press reports in Viennese and Budapest newspapers in December 1946 that Mengele had been captured and was in jail. At last, here was the chance Perl had waited for to denounce the monster. In January 1947, she wrote to the Washington office in charge of the war-crimes trials to offer herself as a witness. In her letter, Perl expressed her desire to testify "against this most perverse mass murderer of the twentieth century."

Perl didn't even receive the courtesy of a reply.

She wrote to them yet again in the fall of 1947, renewing her offer. "I have learned the trial of the greatest 'mass murderer' Doctor Josef Mengerle [sic] will be held in Nuremberg . . . I would be very pleased to go there as a witness . . . and awaken the conscience of the world."

Dr. Perl would certainly have made a dramatic witness at Mengele's trial. After months of working side by side with him, the articulate young woman could have described the enormity of his crimes. Even if he had been tried in absentia, testimony by Perl would have caused such a storm that a massive manhunt might well have been launched. But the sheer incompetence—or duplicity—of U.S. authorities led them to overlook both Perl's letters and her extraordinary book. By the fall of 1947, there was only a halfhearted effort to pursue Nazi war criminals, and Perl's second offer was lost in the bureaucratic shuffle. A review of the U.S. Army's file on Mengele shows that her October 1947 letter was stamped "Hold for Final Action." A small note scribbled next to Mengele's name states quite erroneously that he had been "tried by Poles."

America's Nuremberg team made at least three major mistakes with respect to Perl, all of which helped Mengele evade capture. When reports of Mengele's arrest proved to be false, they didn't bother to reply to her first letter and inform her that the war criminal was not in their custody. By the time of her second letter, they wrongly assumed he was in the hands of the Polish government. They compounded the error by concluding that Poland had already tried and sentenced Mengele.

But it was the Nuremberg team's last mistake which guaranteed that SS Dr. Josef Mengele's trial would never, ever take place. Perl's October 1947 letter was passed around from office to office, from bureaucrat to bureaucrat, in Washington. Then, it was dispatched overseas to the men responsible for the Nuremberg prosecutions. By December, it had finally reached the desk of General Telford Taylor, chief counsel for war crimes and the lead prosecutor at Nuremberg. In January 1948, General Taylor wrote to Perl, apologizing for the delay in responding. Taylor's brief letter ended with a most startling assertion: "We wish to advise our records show Dr. Mengerle [sic] is dead as of October, 1946."

No one knows what prompted the Nuremberg investigators to conclude that Mengele had died—in 1946, no less. Even an exhaustive Justice Department inquiry more than forty years later failed to clarify what happened. But all the available evidence points to the Mengele family. Beginning with Irene's clever donning of mourning clothes and her recital of a mass in memory of her "late" husband, the Mengele clan's strenuous efforts at spreading disinformation had obviously succeeded. Their line that Josef was missing and presumed dead on the Eastern front had somehow found its way even into official American war-crimes files. Since Nuremberg officials believed Mengele was dead, they failed to continue the hunt for him. The only other possible conclusion is that U.S. authorities deliberately helped the Angel of Death evade capture, but nothing has surfaced to buttress such a theory.

In Mangolding, Mengele fretted over problems more pressing to him than being tried for war crimes. His writings on the period suggest that his daily concerns centered less on avoiding capture than on forestalling a divorce from his unhappy wife. He remained deeply upset over his crumbling marriage. He continued to mourn his mother's passing and the Günzburg life he would never again know.

VERA GROSSMAN:

The Communists made life very unsettled for us.

One day, a rabbi from England came to visit my parents. His name was Dr. Shlomo Schoenfeld, and he ran a famous school in England.

Rabbi Schoenfeld was traveling through Eastern Europe, collecting

Jewish children to take back with him to Britain. He wanted to offer them a better life.

He convinced my parents that it was not safe for Olga and me to remain with them in Eastern Europe. We were nine years old at the time. It was only two years after Auschwitz.

Mother sadly agreed. She hated to part with us. She always called us her "miracle twins" because, as she liked to say, it was a "miracle" that two little girls had survived both Auschwitz and Dr. Mengele.

I understood her decision. I may have been nine years old, but felt like a grown-up person.

Still, I was heartbroken to leave her.

I remember the day they took us to have our pictures taken for our travel documents. I refused to smile.

We were placed on board a train, this time bound for England. We were with lots of other war children. Olga and I celebrated our tenth birthday on that trip.

I felt very angry.

6

THE STORY OF ANDREAS

TWINS' FATHER:

By the late 1940s, most of the surviving Jews of Eastern Europe—not just the twins—had realized they had no future there, and were trying desperately to get out. We all felt home was in Palestine.

But it was almost impossible to emigrate—at least legally. Many members of my family, including my twin, Magda, managed to slip in illegally, through Cyprus.

My wife and and I wanted to get legal visas. But it was extremely difficult to do so. The Communists didn't want to let us leave.

JUDITH YAGUDAH:

I kept pushing my mother to leave. I belonged to a Jewish Zionist movement, B'nai Akivah, which exerted a great deal of influence over me. The group taught me to reject Communism and embrace Zionism. They made me long to go to Palestine.

This was a trend among all the Jews. There was always that lingering fear—that they would once again persecute the Jews.

We didn't want what had happened once to happen again.

Three years after the end of the war, the man who official Nuremberg records claimed had been dead for two years was still living at the Fischer farm. Irene, his discontented but dutiful wife, continued to act the part of the grieving widow. One day, she even had a funeral mass said in memory of her "late" husband at the local church. But as the terrible decade came to a close, it was becoming more and more difficult for Mengele's wife to rationalize remaining in her in-laws' insulated Bavarian town. Germany was beginning to come out of its own postwar mourning period. Lights were turned on at night for the first time in years. There were dances and parties. Irene saw her youth slipping away, and her marriage increasingly appeared to have been a foolish and unrewarding venture.

The tension and acrimony in Irene's marriage to Josef dragged on. They continued to meet occasionally, but reunions were marred by bitter arguments and recriminations, smoothed over by temporary reconciliations. Both were coming to the realization that they didn't really know each other. Mengele's writings suggest he perceived Irene as superficial, a self-centered "egotist" who was more concerned about her own pleasure and convenience than his immense suffering. He seems to have deeply resented her complaints about her life without a man in the house. Irene, on the other hand, saw him as irritable and domineering, as always trying to tell her what to do with her life and their son. He could also be jealous and possessive; as his book shows, he wove paranoid scenarios of Irene's supposed liaisons, inventing lovers and trysts. Again and again, he questioned her about her relations with other men. The most innocuous encounter she might tell him about was proof of an adulterous relationship. There were instances when his rage was so overwhelming, she thought him unbalanced.

The obsessive qualities Mengele had exhibited throughout his adult life had no release at the farm. In medical school, and later at Auschwitz, Mengele was able to throw himself into his work, becoming engrossed in the most minute details of a test or an experiment. His fanaticism, the long hours he spent at the laboratory, and the fastidious way he performed his work, had earned him praise and admiration. But now, there was only his marriage to occupy his troubled mind.

It is by way of his autobiographical "novel" that Mengele provides the most truthful insights into his disintegrating relationship with Irene. While hiding behind a fictional protagonist, Andreas, Mengele says much about himself as a husband, a lover, and a man.

In the novel, Andreas, a man in hiding, is driven to the verge of madness when he learns that his wife, Irmgard, is going on holiday to a Bavarian resort. Convinced she is planning to meet her lover, he pleads with her not to go. When he hears she has gone despite his entreaties, he decides he must join her there. Friends try to dissuade him from undertaking the foolish journey. They warn him of the extraordinary risks he is taking; he is, after all, a wanted man. Crazed with jealousy, Andreas goes anyway, determined to save his marriage, "this last remnant of my crumbling life." After a long, tiresome train journey, during which he reflects on his failed marriage, the hero goes to the inn where Irmgard is staying. She has already left by the time he gets there, but he continues the chase in the hope of catching her with her lover. But when Andreas does meet up with her, Irmgard is alone. There is no other man.

In real life, Mengele's extreme jealousy appears to have extinguished whatever feeling remained between him and Irene. For Mengele, it had been a marriage of vanity. He had seen in the tall, lissome blonde the female embodiment of the Aryan ideal. Irene had been much more romantic. She had fallen deeply in love with the handsome Josef. But now she longed for a conventional life with a home, children, and a husband she could depend on to be there. "I am not an old soldier's wife—almost all the men in our circle of acquaintances are back home again. But you cannot come home," Irene's alter ego, Irmgard, says at one point.

In the novel, the alpine reunion proves to be a complete letdown. Although Andreas and Irmgard do spend the night together, it is "disappointing." Mengele the novelist spares us the details, saying only, "The night brought the realization he [Andreas] had long suspected but had not wished to accept: the marriage was over."

Elsewhere, Andreas sadly observes that "with the Third Reich, my marriage also ended." What Josef failed to see was that not only his marriage but also the life he had dreamed of effectively ended with the fall of Hitler. The Mengele family remained loyal to Josef and to his version of events. Mengele's father and brothers steadfastly refused to believe the stories that were being told about their beloved Beppo. They closed ranks behind Mengele out of a combination of familial duty and genuine affection for the man they had known only as genial, carefree, and utterly endearing. But his family's loyalty and concern could only help Mengele survive in obscurity. His hopes for a brilliant future were dashed.

In a bizarre fashion, Auschwitz had damned Dr. Mengele almost as much as it had the poor Jews he had slaughtered. Mingled with the remains of the innocent men, women, and children he had dispatched to the crematorium were the ashes of his own ambitions. In destroying them, he had destroyed himself. Yet even now, he did not realize that he would never lead the honorable life he had hoped for, never enjoy the respect he believed he deserved.

He could not see that his life was over: that the postwar world would have no place for old die-hard Nazis. Although he was a war criminal wanted by several countries for mass murder, in between his bouts of despair, he kept hoping he would be called back into the world of academia. It is ironic that this devotee of Darwin failed to heed the master's central lesson: Only those who adapt to change survive.

The years at the Fischer farm passed slowly. Mengele's tedium was relieved only by the visits of family members, or furtive trips to Autenried to see his wife and young son. Lonely and restless, Mengele decided he was tired of being a farmhand and fugitive, and that he needed a change of life. He had heard that Nazis were honored and respected citizens in some South American countries, especially Argentina. The Argentine strongman Juan Perón was known to be a great admirer of Hitler's Reich. A new future beckoned.

ZVI THE SAILOR:

An underground Jewish organization arranged for my twin brother and me to escape to Palestine. We sailed on a ship manned by Jewish soldiers of the British Army. It was a very small ship, and there were hundreds of Jews on it. The British did not want any more Jewish refugees in Palestine. Our boat was stuck in the port of Haifa for two months. Finally, they let us disembark. It was the last ship of Jews the British allowed into Palestine before they began sending the refugees to Cyprus instead.

❧

ALEX DEKEL:

I learned that we had been intercepted by two British warships that insisted upon escorting their ship to Cyprus. If the captain refused, they threatened to open fire. In Cyprus, we were treated decently, fed, and given medical attention. One day, a young man, hardly much older than I, spoke to me and told me he was a member of the Haganah. He whispered that I should be prepared to escape from Cyprus that night. The Haganah had been secretly smuggling physicians and young men out of the camps and smuggling them into Palestine. Around 3:00 A.M. he woke me and whispered that we should now make a run for it; we had to reach a motor launch that lay off the beach about a mile down the coastline from the encampment. I was to leave my belongings and never look back. I ran for my life in that darkness of the early morning.

But there was no gunfire. The British didn't know of our plans and departure. I slipped through an opening that had been cut through the wire fencing surrounding the camp. It reminded me of the other fences—electrified—at Auschwitz. I ran along the beach and tried not to remember the many people I had seen deliberately reaching for that other camp's fence, instantly ending their existence.

Some miles off the coast of Palestine, all of us transferred to a flotilla of small rubber rafts, to escape detection by the British. We rowed these to within half a mile of land, took off our clothes, and swam the rest of the way. Exhausted, I collapsed onto the beach without papers, belongings, and clad only in soaked undershorts. But I was home.

I was officially accepted into the Haganah, and it was this event that made my survival at Auschwitz worthwhile. Now I had a purpose in life again; a State to be settled, fought for, cherished.

❧

EVA MOZES:

In May 1948, Israel became a country. What a wonderful feeling! I said to myself, "I wish my parents could have lived to see this."

A Jewish state had only been a vague dream for them. My father had been an ardent Zionist, but he had talked about a Jewish homeland as if it were a fantasy that could never come true.

At the same time, life in Romania was becoming so difficult, we decided to try to emigrate to Israel. We applied for visas.

But the Israeli government would only grant my sister and [me] the necessary papers—they didn't want to give visas to my aging aunt and uncle. And the Romanian government didn't want to let Miriam and [me] out—although they were perfectly prepared to let my aunt and uncle leave. Young people are always in demand. Old people, nobody wants.

But Miriam and I did not want to be separated from our only living relatives. By now, we considered ourselves a family.

❧

LEA LORINCZI:

The Communists did not want me to leave Romania. They called me to their party office and said, "Why do you want to go to Palestine? You can stay here and you can be anything you want to be."

They promised me the world—a brilliant future under the Communist leadership. But my father, his new bride, and my twin brother, Menashe, were determined to move to Israel.

I told the Communists, "After what I went through, I would never live apart from my family—never, never."

Although his marriage had collapsed, Mengele still expected his wife and son to accompany him to South America. The man who had always been obsessed with the trappings of success, with style over substance, evidently could not bear to think he had failed as a husband. He chose instead to speculate that Irene secretly believed the stories circulating about his actions in the war. "There are people in her circle of acquaintances who wish to convince her of my guilt," Andreas says of his wife in the fictionalized memoirs. "We must get through the present and start over abroad."

But Irene was apparently motivated by other considerations than whether or not her husband was a mass murderer. As much as she hated Günzburg, she knew Argentina would have even less to offer. For the cosmopolitan Irene, who in her youth had treasured visits to Paris and Florence, the notion of residing so far away from her beloved Europe was simply unthinkable. The professor's daughter craved a cultured society she doubted would be available to her in South America. She categorically refused to accompany her husband, and began instead to plan her own return to her native Freiburg.

Mengele was floored by Irene's adamant refusal to move, and tried repeatedly to convince her to change her mind. In the novel, Andreas expressed shock and dismay over Irmgard's insistent demands for a divorce that would allow her to rebuild her life. "Man, woman, and child are a unit that one must not destroy," he sternly observes. The admonition falls on deaf ears—Irene's as well as Irmgard's.

But even the prospect of a solitary life was not enough to deter Mengele from leaving Germany. He missed his work and hoped that in Argentina he could resume his scientific career—perhaps even continue the research on twins. He had apparently not forgotten his children of Auschwitz. His first concern was getting out of the country without being caught.

PETER SOMOGYI:

As we became more involved in the Zionist movement, my brother and I decided we had to move to Israel. Since there was no way of leaving Hungary legally, we decided to escape. We would sneak out with our Zionist group.

To plan his escape, Mengele turned as always to his family for help. He bid good-bye to the Fischers, thanking them for their hospitality these many years. At least one member of the family was deeply relieved. According to an interview he gave years after the war had ended, Alois, the farmer's brother who had never liked or trusted Mengele, was glad the Nazi was going to look for another hideout.

In 1948, Mengele returned to Günzburg to plan his getaway. But instead of being allowed to live in the family villa, he was reduced to spending several months hiding out in the woods near the old town. His family gave him whatever he needed to be comfortable, even as they arranged his safe exit from Germany. Although they loved their Beppo, the Mengeles saw the need of getting him speedily out of the country. They, too, could do without the tension and danger of harboring a war criminal.

Mengele needed identity papers, yet despite their wealth and influence, the Mengeles didn't have easy access to forged documents. Irene finally managed to get her husband a passport on the black market, but it was of such poor quality that it was quickly discarded. Mengele had no choice but to try to undertake the journey without any papers.

PETER SOMOGYI:

My twin and I fled Hungary with the Zionist movement on April 4, 1949.

Our flight was very well organized. There were about ten young people in our group, as well as a few parents. We had guides at every stage of our journey. These guides had been well-paid. In those days, there were a lot of people who made their living sneaking people out of Europe.

The border guards had been bribed, so we were able to cross without any problems. But instead of going directly south, we went up north through Czechoslovakia, and then on to Vienna.

In Vienna, we had to bribe Russian soldiers to pass through. The city was then under Russian control. We stayed in an old schoolhouse, where we were joined by scores of other Jews trying to flee.

We spent three weeks in Vienna, and then made our way to Salzburg. In Salzburg, we stayed in an old concentration camp, I don't know which one. I didn't care—and surprisingly, I didn't feel particularly bothered by that. We didn't spend too much time there.

After about a week, we went to Bari, in southern Italy. There, it was very easy to move around: Everyone accepted bribes. They let us go through even though not one of us had passports.

We waited for the boat that was to take us to Israel. There were hundreds of us by then, gathered from every corner of Europe, awaiting passage to the Promised Land.

We were all very excited. We felt for the first time that we didn't have to look behind our back and hear someone saying "dirty Jew."

Mengele hired several guides to help him with his escape. He began by taking a train to Innsbruck. Once there, he went to the Brenner Pass, a route that led into Italy through the Alps. He reached the border that divides Austria and Italy on Easter Sunday, 1949. A guide helped him across the mountainous terrain, and the two made the crossing in the dead of night. In Mengele's thinly disguised account of his flight from Germany, he describes the difficulty of the crossing. His paid companion, noticing he is huffing and puffing, asks if he is strong enough to continue on foot. The hero, somewhat mortified, replies that he is in "top physical shape."

Mengele's novel provides a surreal, even poetic rendition of the flight, from a description of the moon in its first quarter to the edelweiss and primroses dotting the Alps. Throughout his account of Andreas's life, Mengele veers from the dry and factual rendition of events to long passages in which the hero's sensitive and artistic sensibilities are paraded. Even at Auschwitz, the SS doctor had loved to show off his cultured side, whistling the tunes of the great composers as he sent inmates to the gas chambers. Perhaps to refute his monstrous postwar image, Mengele's writings always highlighted his appreciation of beauty, music, and art. It is almost as if he anticipated his reader-critics of the future, and wanted to reveal his "true" nature.

Once Andreas is near the border, his guide leaves him to fend for himself. He continues walking until he reaches Italy, and boards a train to Sterzing, where another guide is to meet him. With time on his hands, he checks into the Golden Cross Inn and nervously awaits his contact. He keeps glancing around for any suspicious characters who might be tailing him, and is especially afraid of running into Americans. But everything goes as planned, and an Italian with the code name "Nino" arrives to meet him.

The fugitive spends approximately one month at the Golden Cross Inn. Yet another contact comes to take his photograph, and returns with an expertly forged German ID card. Armed with this document, Andreas can now proceed safely to his next stop, Genoa, where he will try to get the additional papers he needs to leave Europe.

In writing the "novel" of his escape, Mengele certainly had literary pretensions, but the only evidence of artfulness would appear to be his change of place names and individuals. Historians who have tracked the war criminal's path out of Europe found that Mengele did indeed acquire an ID card from a town in the South Tyrol called Taormino. The document was made out to "Helmut Gregor," a pseudonym Mengele would use for years to come.

The Tyrol, located between Austria and Italy, had a heavily Germanic population. It had long been a point of contention between Italy and Germany. When Italy fell to the Nazis in 1943, the Germans moved in and promptly issued their own identity cards. But after the war, inhabitants of the Tyrol were designated as "stateless" by the International Red Cross because of the unresolved dispute over sovereignty. The confusion was beneficial to Mengele, who was able to obtain an International Red Cross ID card very easily. Thanks to this card, he would later be able to obtain the Swiss passport that would enable him to leave Europe and enter South America safely and legally.

HEDVAH AND LEAH STERN:

When we left Eastern Europe for Palestine, we went from one war to another war, from the Nazis and the Communists straight into the arms of the Arabs.

We sailed to Haifa from Marseilles in 1948. But when we arrived, we couldn't get in because of the war. And so we were forced to sail all around Egypt, to Alexandria and Cairo.

In Alexandria, the Arabs learned there were Jews on the boat. They wanted to board and kill us all. We had to stay locked up inside our cabins for fear of being slaughtered by the Egyptians. The captain told all the passengers that as long as we stayed locked up, he would protect us. But if we dared to venture out—even to the deck—he could not be responsible for our safety.

We continued sailing around the Levant. We went to Beirut, back to Egypt, and finally to Haifa, which allowed us to enter.

We had expected to find the "land of milk and honey,"—but there was no milk and no honey, only more war.

We were placed in an orphanage for girls in Tel Aviv, where they found us jobs in a factory. The fighting made us very nervous. There were no shelters. These were hard times.

We were happy to be in Palestine, of course. But together with the happiness was a deep sense of sadness, of mourning for everyone who wasn't here with us.

We were alone, the two of us, in a strange country. We felt very broken down. We longed for a family.

Before Mengele left Sterzing for Genoa, his old Günzburg chum Hans Sedlmeier came to see him with news from the family. Karl Mengele, who had finally been released from the internment camp, sent money to help his prodigal son get through the difficult weeks and months ahead. Mengele Sr. still doted on his oldest boy. He didn't want his son to lack for anything while on the run. And throughout the years to come, it would be Josef's father—using Sedlmeier as a preferred intermediary—who saw to it that the war criminal was well-cared-for in his exile.

A benevolent, even humanitarian, streak was surfacing in Karl Mengele. Perhaps it had always been there. For not only did he take care of Josef, he also became Günzburg's favorite and most generous philanthropist. After the death of his parsimonious wife, Karl Sr. sponsored many charity fund-raisers. He got into the habit of placing large sausages in the windows of the homes of Günzburg's impoverished residents. And long after World War II, when the Mengele name had come to symbolize evil the world over, the death-camp doctor's father was the most venerated and venerable man in his town.

That he was Günzburg's largest employer certainly did not hurt his high standing. But Karl Mengele clearly possessed a munificent spirit. It harkened back to the early years when he had sought to give his sons cars and fine clothes over his wife's objections and contrary to her penurious inclinations. In the past, he had had to keep his generosity in check to avoid enraging Walburga. But now, she was no longer around to raise a fuss. More important, the elder Mengele shrewdly recognized how useful it was to appear civic-minded, if only to counter the barrage of negative publicity about Josef that persisted through the years.

Although Sedlmeier had been Josef's childhood friend and schoolmate, it was out of respect for the elder Mengele that he devoted so much of his energy to aiding his war-criminal son. Years after Karl Sr. died, Sedlmeier continued to keep a watchful eye on Josef, visiting him regularly in his South American hideouts and making sure he had money to live on.

For their Italian reunion at the inn, Sedlmeier brought, in addition to cash and family greetings, another wonderful surprise: a package containing scientific slides Mengele had gathered at Auschwitz and sent home for safekeeping. Mengele was delighted. They would be useful in the laboratory work he hoped to resume in Argentina.

Carrying the money and his precious box of slides, Mengele proceeded to Genoa, where he was met by yet another guide—"Kurt" in the novel. Plump but agile, Kurt was a nervous creature who quickly tired of the hero's haughty airs and prima donna attitude. At one point, Andreas complains that he has not yet explored the lovely Mediterranean city, or even seen the "*Mare Nostrum*," as he pretentiously calls the sea. Kurt is not impressed by the hero's command of Latin. "You are not exactly a harmless tourist," he snaps at his erudite charge.

EVA MOZES:

Before we left Romania, Miriam and I wanted to go back to our native village and visit the cemetery where our grandparents were buried. But the villagers were very hostile. They were attacking former landowners—and our family had been one of the wealthiest in the town. I could not even go back to my village to see it one last time.

In Genoa, the first order of business was to get Mengele a fake passport from the Swiss consulate. The fact that he already had an International Red Cross ID made this very easy. The ID card was viewed as a legitimate document by the Swiss government, which used it as a basis for giving war refugees legal passports. Using this route, many Nazi war criminals were able to exploit the humanitarian instincts of the International Red Cross and Switzerland to obtain the necessary documents to flee Europe. Of course, neither entity knowingly helped Nazis evade justice.

At the Swiss consulate, Mengele was certain the attractive wide-eyed clerk was staring at him: Did she recognize him? Although the woman's demeanor was professional and she seemed eager to help, Mengele feared she had guessed his true identity. Occasionally, she would look up from her papers and smile at him. It was an innocent gesture, and quite possibly due to the fatal attraction Mengele always held for women. But it was enough to make the former death-camp doctor lose the poise and proud manner that had so irritated his Italian guide. Mengele now sought to get the formalities over with as quickly as possible.

His worst suspicions seemed confirmed the next day, when Mengele discovered the passport she had issued him was useless. Purposely or unwittingly, the smiling clerk had stamped the date he applied for the papers on the line marked "expiration date." This meant the passport had expired the previous day. Mengele was sure it was a trick. But he went back to the consulate and was relieved when another clerk willingly issued him a valid passport without any questions.

The memory of the consulate clerk haunted Mengele for years thereafter. He kept seeing her vague smile, her eyes that seemed to see right through him. Over a quarter of a century later, he was able to vividly conjure her for his novel.

Next came the mandatory physical examination at the harbor. A Croatian doctor performed the examinations and administered the required vaccines. For a few extra lire, the doctor obligingly back-dated the certificate of innoculation by two weeks, to enable Mengele to obtain an exit visa. Mengele had to undergo another physical exam at the port. There, he noted the disgusting habits of his professional colleague. The Croatian doctor used the same instruments again and again, paying no heed to the dangers of transmitting infection. The veteran of Nazi death-camp infirmaries, where knives were used instead of scalpels and patients were left to shiver and die on bare planks of wood, was appalled at the lack of sanitation at the port of Genoa.

The last item Mengele needed to escape was an exit visa. Mengele's Genoan contact had planned to bring his application to the attention of an Italian official known for his kindness toward refugees—especially those who exchanged money for his favors. But by a twist of fate, the man was away on holiday. Mengele's ship, however, was set to sail in just three days. In the novel, the vessel is called the *North Queen*, startlingly close to the actual name of the ship Mengele boarded: the *North King*.

JUDITH YAGUDAH:

It was the great exodus out of Europe.

Mother and I finally got our visas. We took a train to Constanza, a port on the Black Sea, and then we boarded a ship.

Thousands and thousands of Jews were using the same route.

❧

EVA MOZES:

It took nearly two years, but at last, all of us—my twin sister, my aunt, my uncle, and I—finally received our visas. We quickly made arrangements to sail to Palestine from Constanza.

We started packing furiously. Tables, beds, clothes! We were going to wrap everything we owned and take it with us.

But then we were told we could take only fifty kilos of belongings. Everything else was to be signed over to the Romanian government. And then, when we got to the ship, we learned we would only be able to board with the clothes on our backs. Our ship was built to hold a thousand people—instead, three thousand were to be crammed in.

I wore three dresses, one on top of the other. Miriam, too.

We had come back from the camps with nothing, four years before, and now we were leaving with nothing.

In the novel, Andreas awkwardly tries to slip the official in charge of granting exit visas twenty thousand lire. But the functionary will not be bribed, and starts questioning Andreas closely. When Andreas tells him he is from a small town in the South Tyrol, he is openly skeptical. With a flick of the wrist, the official dispatches Andreas to jail.

In real life, Josef Mengele did spend several weeks inside a Genoa prison—the only punishment he ever received for his murderous deeds. The man who once condemned thousands to a fate far worse than an Italian prison cell felt he would go mad. He paced up and down, raging like a wild animal, according to his novel. Gone was the controlled camp doctor who calmly went about his duties sending people to die with a smile. At no point did it occur to him that his imprisonment might be a punishment for his Auschwitz crimes. Indignation, not remorse, was Dr. Mengele's only emotion.

In an ironic turn of events, several of Mengele's prison peers were drug addicts and cripples and a host of other "inferior beings" whom he would have swiftly dispatched to die in the crematoriums—or to be tortured in his laboratory—in former years. There was a dwarflike, handicapped street musician and a doctor who was a morphine addict. When the doctor began showing withdrawal symptoms, shaking and crying until he cut himself breaking a window. Mengele watched and didn't even try to help. The addict was simply another defective human being responsible for his plight, not worthy of care or compassion.

The weeks Mengele remained in jail seemed an eternity. But at last, the friendly Italian bureaucrat his contact had tried to reach earlier returned from his holiday. Apprised of the situation, he quickly set Mengele free—and even apologized to the war criminal. Mengele was able to board the ship whose sailing had somehow been miraculously delayed. Thanks to the corrupt official, his ticket was even upgraded from tourist to second class: This was just a small way the Italian government showed how sorry it was for Mengele's unfortunate detention.

Mengele's escape from Europe belies all the exotic theories that existed for years on how the Angel of Death had eluded capture. There was no sophisticated Nazi network in operation to ferret him out to safety. He was not aided by the American government or any of its intelligence agencies. The Vatican had no apparent role in his flight. Rather, like so many other war criminals, Mengele simply paid a series of accomplices with no particular affiliation and took advantage of the International Red Cross.

In chaotic postwar Europe, with thousands of refugees and displaced persons in need of false papers, there were many men like Nino and Kurt around willing to help—for a fee. Even the high-level Italian official who released Mengele from jail was probably not linked to any nefarious underground Nazi brotherhood. Rather, he was simply used to getting paid generously for his services. Ultimately, petty corruption inside the Italian bureaucracy was what saved Dr. Mengele, even after he'd finally been caught and placed behind bars.

By the time Mengele set sail for Argentina, he was no longer quite as confident as when he'd first made the decision to leave Germany. In the course of the long transatlantic voyage, he learned it was not going to be easy to start anew. His German credentials were not enough for him to practice medicine in Argentina: He would need to get recertified, a process that meant going through school again, possibly for many years.

Yet when the *North King* sailed into the bustling harbor of Buenos Aires, Mengele felt some of his old optimism and boyish excitement return. It was that feeling of endless possibilities that had graced his youth in Günzburg, rekindled by the charm of the new land, a sense of hope as heady as the scent of a summer night on the Plaza.

EVA MOZES:

It was early in the morning when our ship approached Haifa.

We watched the sun rise over Mount Carmel. It was one of the most beautiful sights I had ever seen.

Most everyone on the boat was a Holocaust survivor. We all stood up and started singing "Hatikvah," the Jewish national anthem.

We hugged and kissed each other. We felt at last we had come home.

FUGITIVE'S IDYLL

VERA GROSSMAN:

After leaving our parents, Olga and I were taken to a convalescent home in Ireland. It was located in an old castle, not far from Dublin. It was a beautiful castle—really beautiful—just like in the movies. It even had a moat.

There were fields all around, miles and miles of green fields. I remember picking wild apples from trees, and strawberries from strawberry fields. I had never tasted strawberries before.

They fed us constantly. After the war, my sister and I were skinny and undernourished. We had problems with our lungs. I became very friendly with the cook. She was a big fat woman, and I remember hugging her, and clinging to her apron.

I loved being inside the kitchen, loved the way it smelled. All my life I had been hungry, and for once I was getting more than enough to eat.

Many of the other children there were Holocaust survivors. Some had spent the war in hiding. Olga and I were quite popular. We were nicknamed "the twins."

I got a reputation as a real mischief-maker. I led the other children. At night, we would put sheets over our heads and wander around the castle, making believe we were ghosts.

I remember feeling very happy there. It was a carefree, idyllic period.

In Buenos Aires, Mengele discovered a metropolis that was thoroughly Latin American, yet longed to be European. Although smaller than São Paulo and less attractive than Rio de Janeiro, Buenos Aires was unquestionably the capital of South America, the city that had the most to offer.

Cafés lined the wide avenues, as in Rome and Paris. Late at night, one could amble through the streets and hear faint strains of jazz or the samba. The cabarets were crowded every night with bejeweled young women and their wealthy escorts.

Reigning over them was the charismatic Juan Perón, a dictator's dictator, with a movie-star smile and an iron-clenched fist, ruthless and charming. Perón was the embodiment of machismo—yet he worshiped only one woman, his wife, Eva.

During the seven years they were married, Evita was seen everywhere, discussing military strategy with the generals, whispering to Juan at state meetings, helping him to make policy decisions, presiding over groundbreaking ceremonies for charitable institutions, and, after a long day, idly smoking a cigarette in a nightclub where she had once been a chorus girl. Perón, shrewd politician that he was, realized that his wife was one of his best assets in retaining control over the fickle Argentinian population. Evita was adored, a mythical creature, enshrined even before her premature death.

Of course, Evita had her critics, including the American diplomats who thought her a hindrance, and maybe even a threat, to Perón's power. But Perón shrewdly disregarded the Americans' opinion. He had his gripes with a country that persisted in calling him a closet Nazi even though, as he never tired of reminding them, in the late 1940s he was welcoming more Jewish immigrants than the United States. As for the Nazis he allowed to immigrate to Argentina, he argued that letting them in was yet another demonstration of his humanitarian bent. Juan Perón would not turn anyone away. Thanks to him, Argentina was a haven for all war refugees, Jewish or Nazi, fleeing Europe in the late 1940s and early 1950s.

Concentration-camp guards and their victims alike were drawn to Argentina for its Western charm and culture and, more significantly, its open-door policy. Nazi war criminals fled their old homes. Jewish survivors sought new homes to replace the ones they had lost. They knew that Perón would not tolerate any displays of anti-Semitism. The two did not mix, although Nazis did occasionally frequent Jewish-owned shops. Both groups discovered large communities of their compatriots who had been in Argentina for years. Even before the war, Jews had flocked to Argentina—one of the few countries that allowed them in—and had built a cohesive community.

Similarly, thousands of Germans had emigrated to Argentina, beginning a century before, in search of better economic opportunities than those available in Europe. They founded German clubs, German schools, German shops, and—most tantalizing for Mengele—a German hospital, which served as a symbol of what his future would hold if he flourished in the New World. And there were also several Nazi organizations that pledged loyalty to the Reich and the Führer even years after the war was over. Endemic to the community was a strong nostalgia, a heartfelt longing for their homeland that underscored every activity.

JUDITH YAGUDAH:

Mother regretted leaving Romania. She disliked Israel from the moment we set foot on the ground: she said it reminded her of Auschwitz.

When we arrived, we were taken to a refugee camp in Atlit. We lived not in houses, but in tents. The country was very poor. Whatever you needed to live—bread, milk, eggs—had to be purchased with coupons. Life was very hard.

In our camp, there were a lot of immigrants from Asia and North Africa. The men walked around all day in their pajamas—striped pajamas. At Auschwitz, male inmates had also worn striped uniforms that looked a lot like these pajamas. That's what reminded Mother of the concentration camp.

"This is like Auschwitz," Mother would say.

❧

MOSHE OFFER:

When I arrived in Israel, all the children I had traveled with had someone they could go to—a cousin, an uncle, a friend. I had nobody.

I was placed in an orphanage. I was the only Holocaust victim there, and it was very hard for me. On the weekends, the children went home to their relatives. I was left all by myself.

I was very jealous of the other children. When the weekend was over, they returned with care packages and pocket money given to them by their families. They could buy themselves little treats, go to the movies. But I had no money at all.

I lived from hand to mouth. Sometimes, I would sneak into the cinema. If I was caught, the ushers or the owner would beat me up and throw me out.

These were very hard times, and I was extremely depressed. I felt completely alone.

Sometimes, they would close the kitchens. Since I always felt hungry, I picked through the garbage for food to eat.

Restaurants in the German-Argentine community disdained the local cuisine, favoring beer over fruit juice and serving schnitzel rather than the ubiquitous steaks. An old Victrola cranked out nostalgic songs from before the war a famous *lied*, perhaps, or a patriotic German anthem. Clusters of old men sat around recalling the days when it seemed certain that Germany would rule the world. Discussions invariably returned to the empire they had lost. Such conversations made these defeated soldiers of the Reich much less gloomy. They spent their days in cozy establishments that bore a striking resemblance to the restaurants and beer halls of Frankfurt, Munich, and Berlin.

For the wealthier, more cosmopolitan Germans, Buenos Aires was a favorite city, a place that offered some relief from the arid drudgery of the South American hinterland. Its natives were also more sophisticated, and there was some social mixing of the Germans with the Argentine upper classes. Wealthy Argentines rivaled the Germans in their tastes and snobbish sense of superiority. There was also a deep and pervasive anti-Semitism among the Argentine elite that was attractive to Nazi immigrants. But as long as Juan Perón was in power, these anti-Semitic sentiments were not allowed to go beyond the private drawing rooms of Argentina's upper crust.

At first, Mengele's world was far removed from Argentine—or German—high society. He arrived in Argentina penniless, the money his father had sent him via Sedlmeier having been used up during his Genoa adventures. A German contact who was supposed to meet Mengele at the boat's landing never showed up.

EVA MOZES:

When we stepped off the boat in Haifa, an uncle was waiting for us. He hugged us. He said he wished our parents had left Cluj and settled in Palestine before the war, as he had done. Then they would have been with us, instead of gassed in a German concentration camp.

But Mother had said no. We were leading a good life in Eastern Europe. She had heard conditions in Palestine were too "primitive," especially for raising young children. She convinced Father to stay put and not make the family emigrate.

Miriam and I cried in our uncle's arms about those lost years and our parents' tragic mistake.

He took us to a distribution center for immigrants. A week later, we were moved to a Youth Aliyah village. This was a center for children who were either orphans or whose parents could not take care of them. These centers had been started in 1934 to rescue children from the Holocaust. After the war, they were used to help children who had survived the concentration camps.

❧

LEA LORINCZI:

When we arrived in Israel, we were met by my future husband—although of course, I did not know then that we would get married. He was my stepmother's brother.

He was very nice. He arranged it so that we did not have to go to any refugee camp. Instead, he got us a room in Jerusalem. My parents, my twin brother, and I all lived in that one room.

Mengele was forced to spend his first few weeks in a cramped room in a fourth-class hotel, one built especially to accommodate the thousands of refugees who poured in each month from Europe. The racial hygienist shared a room with two other people. He made do with a community toilet and sink located at the end of the hall.

Mengele could stroll by the elegant cafés and admire the pretty women. In actuality, he was more impressed by Argentina's soldiers than its women. In letters composed many years later, Mengele bemoaned the younger generation of Germans and what he saw as their "cowardly inferiority complexes." In Argentina, soldiers and generals were revered, and showered with status and privilege. He admired the respect the country gave to "military tradition," and ruminated sadly on the fall of the German Army. He was especially drawn to Juan Perón's imposing personal guards, who stood on duty outside the Casa Rosada, the presidential palace. Despite his poverty, Mengele felt at home—more so than he had in his own country after the war.

EVA MOZES:

I will never forget our first Friday night in the Youth Aliyah village. Miriam and I entered the dining room, and we saw everyone wearing white. There were candles on the table, and wine.

Children from many different countries were seated together—yet they were all speaking one language, Hebrew.

We recited prayers and sang Israeli songs. After the meal, we were taken to a large room where all the young people were dancing Israeli folk dances.

They taught us how to dance. That night, Miriam and I even learned a few words of Hebrew.

I felt so at home. I could have stayed at the Youth Aliyah forever.

Mengele had one contact in Argentina other than the person who had not met him at the dock. The man was also a doctor, named "Schott" in the autobiographical novel, and Mengele went to see him with high hopes. The encounter was disappointing: The doctor was now employed in a weaving mill. The best he could do for Mengele was to get him a job at his company, combing wool.

Mengele realized that he would not get any work commensurate with his experience and abilities. Although the old doctor assured him that many other prominent ex-Nazis were employed in his firm, the prospect of such work was depressing to Mengele, and he decided to forgo the opportunity. He left intensely discouraged.

PETER SOMOGYI:

When we arrived in Israel, it was very hard for any of us to get jobs to support ourselves.

My father, for instance, who had always been very successful running his own business, had a hard time trying to reestablish himself. He simply could not make a living. He was in his late fifties—not a young man anymore.

When I looked for work after getting out of the Israeli Army, I could find nothing. And so I decided to learn a trade. I became an auto mechanic. It was not very Jewish, to learn a trade. We had always been brought up to be "professionals." But I thought it was useful.

I worked as a mechanic for three years. I was very good with my hands.

❧

TWINS' FATHER:

We arrived in Israel on my wife's birthday. Luckily, I was able to quickly find a job in my field, accounting. I worked for a large firm. They let me handle a very glamorous account—one of the largest theaters in Israel. I did well, and they came to rely on me. I kept their books in tip-top order. They would deal only with me.

❧

MOSHE OFFER:

I was fifteen years old, and still living at the orphanage, when I got my first job—as a dishwasher at a nearby restaurant. I had decided I needed to make money. I was going to school, which was expensive. I went to classes in the morning, then I would wash dishes in the afternoon to pay for my studies.

I would come home very late from the restaurant, go to sleep, and wake up early the next morning to go to school.

By chance, Mengele met a carpenter who knew of both a job and a place to live. The carpenter was quitting his job and told Mengele he was welcome to have it. He then directed Mengele to a small rooming house in the Vicente Lopez neighborhood. The Auschwitz doctor would be sharing a small, windowless room with an engineer.

But even that was considerably nicer than the fleabag hotel where he had been living.

In 1949, Mengele began work as a carpenter, presumably a more respectable and interesting line of work than wool-combing. The recipient of a Ph.D. and medical degree discovered he had a knack for building and fixing furniture. Little by little, he started establishing himself. Then, when his roommate's daughter became sick, Mengele was asked to treat her—secretly, of course. Although the little girl lived with her mother, her anxious father brought her to Mengele for expert medical care. Mengele agreed, delighted to be able to put his medical abilities to use.

LEA LORINCZI:

Shortly after we arrived, I decided to study to become a nurse. I had decided I wanted to help the sick. At Auschwitz, a nurse had saved my life. She was a Jewish inmate who worked at the infirmary where they placed me when I got very sick. If it hadn't been for her, I would not have survived the concentration camp.

I was only sixteen years old when I began working as a nurse at Shaare Zedek Hospital, one of Israel's leading medical centers. I had no money. I worked very hard, very long hours. And I didn't speak a word of Hebrew.

One of my duties was to prepare basins of water each morning and wash the patients. One day, an old woman started crying, "Hum, hum, hum." *I had no idea what was bothering her. And then, I realized she was complaining because the water was scorching her.* Hum *was the Hebrew word for hot.*

I began to carry around a little Hebrew-Romanian dictionary in my pocket. Whenever a patient said a word I didn't understand, I simply looked it up. That's how I learned to speak the language.

❧

EVA MOZES:

After a while, Miriam and I were drafted into the army and had to leave the Youth Aliyah village. We were both asked what we wanted to do with our lives.

Miriam immediately asked to serve as a nurse. I wanted to do that, too, but our uncle felt it was not good for both of us to be so alike. He said, "Eva, you are good in math—become a draftsman."

❧

MIRIAM MOZES:

In Europe, my great dream had been to become a doctor. But in Israel, I didn't have the money to study medicine—so I became a nurse instead. I was offered free room and board in the hospital. I paid nothing for my training.

❧

EVA MOZES:

We were separated for the first time at boot camp. Miriam went to the hospital to train to become a nurse. I was sent to work as a secretary. The Israeli government had decided it would not train women to be draftsmen because they would get married and leave their jobs. Instead, I was only trained to type.

I worked in an office in Tel Aviv. Miriam lived with other nurses at her hospital. It was traumatic. We had never been apart before.

❧

MIRIAM MOZES:

It was very hard to live apart from Eva. From Auschwitz until then, I had always had my sister at my side. She took care of me. She was like my mother.

I cried all the time at first. It took a year before I got used to being on my own. I made friends among the nurses. During holidays, Eva came to visit me. We tried to see each other at least once a week.

When I finished my studies, I went to the director of the hospital and told him I wanted to share a room with my twin sister. I told him I had no parents or brothers or sisters except her—she was my only surviving relative. He agreed to let Eva move into the hospital dormitory.

Mengele began corresponding with his family in Günzburg, as well as with his six-year-old son. Irene had finally left the Günzburg area and returned to her beloved Freiburg, taking little Rolf with her. In 1948, she had become friendly with a businessman from Freiburg, Alfons Hackenjos, and the two fell in love. The romantic entanglement was not unexpected: Irene was tired of the war-widow's life. Hackenjos, a prosperous shoe-store owner, offered her the companionship and stability she craved.

Though Mengele's relationship with Irene was over, he had every intention of keeping up with his only child. Mengele's letters to Rolf, who was growing up into an appealing and bright little boy, were always signed "Uncle Fritz." The child had been told early on that his father had died in the Russian Campaign—Rolf had no idea Mengele was alive, let alone that he was one and the same with "Uncle Fritz." The letters make clear that Mengele had lost none of his old flair for captivating small children. He spun colorful tales of gauchos, the South American cowboys, recounting their exploits in raising cattle on the pampas. Sometimes there were amusing drawings, or perhaps a poem or a song composed especially for Rolf. Each time, there was a big stamp, often with the photograph of Evita Perón, that Rolf could add to his burgeoning collection. Argentina assumed mythical proportions in the mind of Mengele's child. It seemed considerably more interesting than his own humdrum Black Forest town. In Argentina, men who were larger than life daily tried to tame the wild beasts that roamed the pampas. And reigning over this magical kingdom were two people—the lovely blond woman of the stamps and his own handsome uncle Fritz. The little boy waited expectantly for the letters from the amusing, glamorous uncle he thought he had never met, who lavished such intense affection on him—nearly enough to make up for the father he had never known.

VERA GROSSMAN:

Parents started coming to Ireland within a few months of our arrival to reclaim their children. Many families were able to obtain visas to emigrate from Eastern Europe to Canada and the United States, and they would stop in Ireland to pick up their daughters. Our group in the castle dwindled to less than twenty girls.

Our own mother and stepfather left Eastern Europe for Israel.

After we had gained sufficient strength, Olga and I and the remaining girls were moved from Ireland to London. We were placed in the house of a rabbi and his wife, who had several children of their own. They owned a large fourteen-room mansion and could take in a lot of us. We attended Avigdor High School, a Jewish school run by Rabbi Schoenfeld, the man who had plucked us out of our homes in Eastern Europe.

School was wonderful, but the family we lived with was cruel to us. They had been paid to take us and the other children in. But they treated us like servants.

Both my sister and I missed our mother very much. And I think the pain of the separation started to show—especially on Olga. Even in Ireland, where we were very well treated, she had begun acting strangely. She talked about people trying to hypnotize her.

We tried our best to keep up the correspondence with our mother. But it was difficult, because as we were becoming more proficient in English, we were forgetting how to speak Czech.

❧

OLGA GROSSMAN:

Because our mother couldn't speak English, we had to write letters to her in Czech. It would take us five hours to compose a single letter.

I remember going to London's West End one day with Vera to purchase a Czech-English dictionary. We wrote our letters word by word. We had both forgotten our native language. We understood it, but we couldn't speak it or write it.

Still, it was very important for us to correspond with our mother.

Mengele faithfully kept up a correspondence with the rest of the family, which was still mourning the untimely death of Karl Jr. in 1949 at the age of thirty-seven, after an illness. Josef's father and Lolo were now running the firm, with the loyal Sedlmeier acting as their right-hand man. Mengele was clearly still loved and remembered by his relatives in Günzburg. They helped him financially as much as they could—his father even sent him some expensive farm machinery to start a business. Josef began selling the equipment, traveling to neighboring Paraguay, where there was a lucrative market. He also invested some of his father's money in a small carpentry business of his own, which flourished as well.

In the tightly knit German community of Buenos Aires, it was not hard for Mengele to meet people and establish his impeccable social —and Nazi—credentials. It was known that he came from a wealthy family in Bavaria that was continuing to supply him with money. Several people knew, and admired, his status as a wanted war criminal who had been a scientist and doctor in the Third Reich. Within a couple of years after his penniless arrival, he was mingling with the expatriate Nazi society and meeting important war criminals like himself.

One of these was Wilhelm Sassens, a Nazi wanted for war crimes in Belgium, who was a journalist for Nazi publications in Argentina. It was Sassens who introduced Mengele to the man who was to become his best friend in exile, Colonel Hans Ulrich Rudel. The World War II flying ace, who had been the most decorated member of the *Luftwaffe*, as well as Hitler's personal pilot, was a devotee of Hitler's ideology long after the war's end. He was also a close friend of Juan Perón, and had helped build up the Argentine Air Force using his expertise gleaned from the war years. Through Rudel, Mengele was finally able to join the elite clique of Germans who enjoyed the confidence of the Latin dictator.

Mengele was predictably drawn to these men, who were, like himself, unrepentant Nazis. Like the old Bourbon kings who had learned nothing and forgotten nothing, they mourned the fall of Hitler and the Third Reich. The specter of a defeated Germany, disgraced in the world, had not dissuaded them from their faith in the righteousness and fundamental superiority of the German people.

During World War II, Argentina had been a center of Nazi activity, giving rise to Allied fears that the Germans might use the country as a base to expand their power to the Western Hemisphere. Indeed, since their settlement in Argentina in the previous century, the Germans had always believed they were biologically and culturally superior to the native population. According to Ronald Newton, a historian who has studied German culture in Argentina, the Germans felt "a communal conviction of superiority to the surrounding culture." The large, solidly entrenched populace was an ideal target for Nazi propaganda; Hitlerian notions of law, order, tradition, and respect for authority appealed to the German expatriates.

Pro-Nazi publications flourished in South America long after the the defeat of the German Army. One of them, the virulently anti-Semitic *Der Weg*, or "The Way," was popular in the early 1950s. Sassens was a regular contributor. In 1953, an article appeared in *Der Weg* that dealt with genetics; it was bylined "G. Helmuth," a transposition of Josef Mengele's alias, "Helmut Gregor." Three years after arriving in Argentina, Mengele still felt vulnerable enough to use a pseudonym of his pseudonym.

The article itself was not remarkable. Intended for the layman, it explained basic principles of genetics, although *Rassenhygien*—racial science—was not mentioned. In fact, it contained no hint of racism, fanaticism, or politics. However, even at the height of the Nazi era, Mengele's published articles had been distinguished by their lack of rhetoric. But because it appeared in the Nazi organ *Der Weg*, Mengele's sober, apolitical piece was, by definition, pro-Nazi.

On the other side of the world, Verschuer was far ahead of his old pupil. He had mastered the scientific lingo of the postwar establishment to perfection and was flourishing. After maintaining a low profile for several years following the 1946 newspaper revelations of his scandalous past, Verschuer had finally obtained an appointment at the University of Münster in 1951. Mengele's mentor was now lecturing before packed classrooms at the university.

VERA GROSSMAN:

Olga began to constantly black out in class. Whenever a teacher asked her a question, she would get panic-stricken and faint.

I managed a little better. I worked very hard. Teachers were always holding up my notebooks and praising me for my neat penmanship. I won a handwriting competition two years in a row.

One day, a geography teacher I liked made fun of one of my notebooks. I'm not sure what was wrong, but he held it up in class and ridiculed it. Some of the other children started laughing.

I got up from my chair and started running out of the classroom. I remember running, running out of school. I was crying hysterically. I ran for miles until I got to a park, and I sat there, under a tree, crying.

A search party from the school found me there. The geography teacher came—he apologized to me.

I think it was the first time I cried. I cried my heart out. I cried for all the years I didn't cry.

The next day, I went back to school. I felt better than before. From then on, I cried more often, and much more naturally. I didn't keep things bottled up inside of me as much.

But Olga had more and more trouble in class. It was very humiliating to her, not to be able to answer the teachers. She was always so weak and pale, so timid. My heart broke for her.

Lost horizons: The doomed family of Eva and Miriam Mozes pose in the garden of their estate in Transylvania in 1943, in what was to be their last portrait together. One year later, the family would be herded onto a cattle train and shipped to Auschwitz. Only the twins—who are in the front row, flanking their mother—survived.

"She asked our mother if she would ever dance again—that was her main concern. . . .": Judith Yagudah is shown here with her twin sister, Ruthie, and their mother, Rosie, just before the outbreak of war. Ruthie, "the livelier twin," could not endure the ravages of Mengele's brutal experiments and perished in an Auschwitz infirmary shortly after liberation. As she lay dying, she kept asking her anguished mother "if she would ever dance again."

"As a young girl, I was much more serious, much more moody than other girls my own age. . . .": Judith and her mother, Rosie, in Hungary, 1946. Shattered by the loss of Ruthie, the two attempted to rebuild their lives as best they could. But Rosie was never quite able to overcome her pain, and remained forever in mourning for her dead child. Judith spent a sullen adolescence, haunted by memories of her twin.

"In the beginning, I thought about my sister, Ruthie, a lot. But then, less and less. Life has to go on, I guess. . . .": A radiant Judith, blossoming into womanhood, is shown with her mother in a photograph taken in Israel in 1950, at the age of sixteen.

"I sailed everywhere—but Auschwitz was always with me, no matter how far I traveled. . . .": Zvi the Sailor in 1959 aboard a ship somewhere on the high seas. The most restless of Mengele's twins, Zvi spent his life sailing here and there, hoping to shed his Auschwitz legacy.

Vera Blau and her twin sister, Rachel, smile as they stand in a street in Israel, shortly after the war.

Olga and Vera Grossman as young schoolgirls in London after the war. Forced to live apart from their mother for many years, the twins suffered acutely, and Olga developed severe psychological symptoms, which eventually required her to be hospitalized for depression.

Samuel and Mordecai Bash, twins who survived Mengele's experiments, shown here in their military gear in Israel after the war. The two brothers, who lost much of their family in Auschwitz, now reside in the ultra-Orthodox section of Israel known as B'nai Brak.

Family album: Peter Somogyi, his twin brother, Thomas, and their sister and parents in Pecz, Hungary, 1936. Somogyi's mother and sister perished in Auschwitz's gas chambers.

"We were Mengele's special protégés—he nicknamed us 'the members of the intelligentsia.' ": Twins Peter and Thomas Somogyi pose for a picture in their hometown in Hungary upon their return from Auschwitz in 1945. The two brothers left Auschwitz with all their early belongings: two kit bags they sewed together using tattered blankets from the concentration camp. During the war, the two brothers had been favorites of Dr. Mengele, who admired them for their cleverness.

Peter and Thomas Somogyi in Trafalgar Square, London, 1956, where they briefly lived before permanently moving to America.

"Dr. Mengele noticed me immediately because I didn't look Jewish.": Alex Dekel with his brother and mother in Cluj, 1941. Although he was not a twin, the handsome young boy impressed Mengele with his "Aryan" good looks, and was selected to take part in genetic experiments. He spent the war in the twins' barracks, undergoing the same painful tests and procedures as the twins endured.

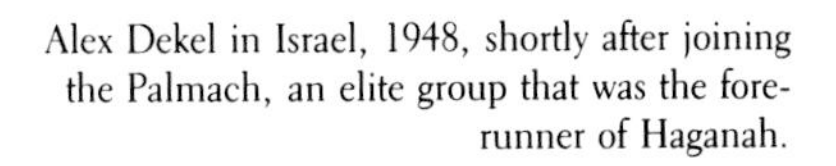

Alex Dekel in Israel, 1948, shortly after joining the Palmach, an elite group that was the forerunner of Haganah.

Menashe and Lea Lorinczi, with their mother, shortly before all three were deported to the death camps. The twins survived, but they never saw their mother again, nor ever found out exactly what became of her.

"Wait for me children, wait for me—we will meet again by the gate.": Leah and Hedvah Stern as little girls in Eastern Europe, in one of the few photographs of their lost childhood they were able to retrieve.

Israel, 1990. The Stern sisters, who live and speak as one, are still inseparable, even in their old age. They continue to wear identical outfits and accessories—even down to their eyeglasses. And although they are now mothers and grandmothers, their thoughts are filled with memories of the mother they lost at Auschwitz.

After the war, Magda Spiegel met up with her husband, and together they sought to start their lives anew. She had another child, to replace the little boy she had lost to the gas chambers of Auschwitz, and ultimately moved to Haifa, Israel. Her son is now a renowned doctor and scientist.

Zvi Spiegel, pictured here with his twin sister, Magda, in Israel, 1984, was known as Twins' Father. Put in charge of the twins' barracks at Auschwitz, he reported directly to Mengele, with whom he forged a good rapport. Because of this relationship, he was able to save many boy twins and to generally make their lives a little easier.

"I remember picking flowers at Auschwitz. . . .": Eva Kupas, whose sole memory of the camp is of the day she went to pick wildflowers, poses with her twin brother in Israel, 1948.

Eva Kupas outside her home on the outskirts of Tel Aviv, 1990.

Miriam and Eva Mozes, Cluj, 1936. Like many of Eastern Europe's Jews, Miriam and Eva grew up in affluent households, where they were doted on and fussed over because they were twins.

Coming of age in Israel: Eva and Miriam, age nineteen, in Israel, 1957. The pretty women thrived on Kibbutz life, where at last they were able to put their awful past behind them.

Return to Auschwitz: During a historic 1985 return visit to Auschwitz, Eva and Miriam Mozes discovered a blowup from the movie made by the Russian army when it liberated the camp. In January 1945, the Russians bundled up the twins in striped camp gear and marched them out of Auschwitz again and again until the scene looked sufficiently "authentic." Here, Eva and Miriam point to themselves in the foreground.

PHOTO COURTESY OF DR. NANCY L. SEGAL

Once installed at Münster, Verschuer set about building the largest institute for genetic studies in West Germany. With each passing year, he felt less cautious about the need for circumspection, and his old ambition led him to push for recognition from his peers. Verschuer was careful to omit any overt reference to racial science from his lectures or publications. In the classroom, he focused on genetics—a field that was eminently respectable.

But Verschuer wasn't entirely discreet. He served for many years on the international editorial board of the *Mankind Quarterly*, a racist and anti-Semitic publication based in Washington, D.C., founded in 1960 by anthropologists who admired the fundamental precepts of Nazi racial science. Verschuer also helped other former colleagues from the Frankfurt Institute attain positions in the mainstream scientific community. His former deputy, Heinrich Schade, taught genetics at the University of Düsseldorf, and also rose in the academic world, although he never attained the status of Verschuer. A die-hard racial scientist long after the fall of the Third Reich, Schade insisted on the need to create a master race of Aryans by getting rid of "inferior" genetic stock. Schade became an adviser and source of inspiration to neo-Nazi movements. Despite these unseemly affiliations, he was respected among colleagues in academia. He and Verschuer were on the best of terms.

The horror of it all was that postwar Germany embraced both Verschuer and Schade with open arms. That one had become relatively discreet, purposefully keeping the lid on his racial theories, while the other remained a flagrant and unabashed preacher on the supremacy of Aryan genetic science, did not matter. The fact that they both thrived is one of the most telling points about the culture. One could soft-pedal one's racial views and publish them only in the obscurity of the *Mankind Quarterly*, a continent away, as Verschuer did, or one could trumpet them from the tallest ivy tower on the campus of the University of Düsseldorf, as Schade did. And still, postwar Germany, which could not have had any illusions about these racial theorists, allowed them to prosper. The point was, it did not matter.

One can imagine the burning envy Mengele must have felt at seeing his old colleagues, one by one, attain the positions he craved. According to Hans Sedlmeier, Mengele made a secret visit to West Germany in the early 1950s and made a side trip to see Verschuer. Perhaps Mengele hoped that his old mentor, who had placed many of his former colleagues in prestigious positions, could do the same for him. There is indeed evidence that even while Mengele was at the Fischer farm in the years immediately following the war, Verschuer tried to get him a university job but was unable to do so because of the furor surrounding his pupil's name.

If, in fact, Mengele and Verschuer renewed their ties with one another, they did so with utmost secrecy. Verschuer's son, Helmut, claimed that his father never once mentioned the notorious young man who had been his protégé for so many years. Mengele also saw a need to protect his old professor: Among the hundreds of letters Mengele left behind at his death that were later released by his son, not one mentions Verschuer.

Aside from the occasional pang of homesickness, Mengele could not complain of his life in Argentina. Although he never penetrated the scientific establishment or practiced medicine again—legally—he did well as a businessman. (There are hints and suggestions he performed illegal abortions while in Argentina, but few details are available.) The sales of Mengele farm equipment, his carpentry workshop, and a generous allowance from his father helped him to lead an increasingly comfortable life. The Nuremberg prosecutors had long since returned home to New York and Washington; several used their prestigious Nazi-hunting credentials to establish lucrative private law practices. The Israelis, involved as they were in building their new country and fighting wars with the Arabs, had yet to send teams of agents to scour South America for those war criminals who had escaped justice. Mengele could live more or less openly: Several people knew of his real identity. His homes became ever more elegant as he moved to more exclusive neighborhoods. By moving, he markedly improved his social status in a city where one's address was a prime indicator of one's station in life. In 1954, the Auschwitz doctor bought a house in the luxurious suburb of Olivos, where many affluent Germans lived. Shortly thereafter, he purchased another, even more splendid villa in the same elegant neighborhood. It was right near Perón's old estate.

JUDITH YAGUDAH:

After three months in the refugee camp, Mother and I were invited to join a kibbutz. But my mother put her foot down. She didn't want a collective life. She wanted to be on her own.

Together, we moved to Netanya. The government helped find us a small apartment. But we had to face the problem of supporting ourselves. Mother had never worked before, and I was still in school.

She began earning money doing the one thing she knew how to do: knitting. She would knit small dresses for babies and little girls. I helped her out. After school, every afternoon, I came home and helped Mother knit the dresses. This was our livelihood.

❧

TWINS' FATHER:

I did so well handling the theater's account that one day its executives asked me to come and work for them as their personal accountant. It was a wonderful opportunity—more money, and the chance to work for a very prestigious outfit.

The theater had three divisions—creative, technical, and financial. I was put in charge of their financial section, handling all their accounting.

I got to see all the new plays—that was a great perquisite of the job.

I tried to forget the Holocaust. I kept telling myself how lucky I was.

As Mengele eased into middle age, he seemed to return to the bon vivant inclinations of his privileged youth. His clothes were stylishly tailored, his friends were at the top of Argentine society, and his income was enough to permit trips throughout the continent and beyond. He was overjoyed when he got a driver's license, and bought himself a sporty new Borgward "Isabella."

The pictures Mengele sent home to Günzburg and to Rolf showed a dapper gentleman, no longer young, to be sure, but still handsome. He was quite fit and trim. There was a jauntiness about him that neither time nor the ravages of the war and postwar years had quite managed to erase. The old *sprezzatura* was there, too, evident in the easy, relaxed way in which he posed alongside his shiny new car.

In 1954, Mengele's father came to Buenos Aires to visit his son. The reunion was warm, and the elder Mengele gave Josef money to maintain his life-style. Karl Sr. would eventually give his son a million marks to purchase half the shares of Fadro Farm, a pharmaceutical company.

The only bad news Karl brought was that Irene was insisting on a divorce; the couple had never obtained a legal separation, and now she wanted to marry Alfons Hackenjos, the staid but kindly Freiburg businessman she had met years earlier. Since the time for manhunts in Germany had long since ended, Mengele could no longer argue that the family needed to maintain the illusion that he was "dead." He had no choice but to agree to have the family lawyers back home prepare the divorce papers. The couple was divorced that year.

LEA LORINCZI:

Even while I worked as a nurse, my family kept talking about shiddochim *—matches—for me. They were always asking me if I wanted to get married. But I did not want to go out with the men they proposed for me.*

I was looking for someone grown-up, whom I could talk to, who would understand me. I needed a husband who would be a friend, a father, a mother. I did not like the boys my own age.

That's how I became interested in my stepmother's brother. He was much older than I was, but I felt I could talk to him. He asked my father for my hand, and I accepted.

Although he was still an exile, banished from his home and unable to resume his profession, his family's concern and generosity helped make the 1950s decidedly good years for Dr. Mengele. But when Juan Perón was ousted in 1955, Nazis as well as Jews panicked: He had protected both. The military coup that threw the dictator out of office after more than twenty years of rule created havoc in the country. Sure enough, after Perón was gone, some of the Nazi publications, including *Der Weg,* were banned. Some war criminals felt concerned enough to flee to other countries. Egypt, where dictator Gamal Abdel Nasser had consolidated his hold, was the favorite refuge.

Despite the coup, Mengele felt secure enough to stay on in his adopted homeland. Even without the protection Perón had extended to Nazi fugitives, Mengele believed no one would come after him in Argentina; he had nothing to fear now.

His belief that he was safe at last was in many ways warranted. A decade had passed since the Nuremberg trials, and the hunt for Nazi war criminals had effectively ended, in Germany as in the rest of the world. In fact, many former Nazis were now flourishing in West German society. Some of the worst offenders of the Third Reich were leading comfortable lives, and, indeed, had achieved positions of prominence in the "new" Germany. Others, like Mengele, had made new lives for themselves and their children in South America.

EVA MOZES:

For years, Miriam and I struggled and had very little money. We had to work a whole month simply to buy one outfit. We shared clothes to stretch our wardrobes.

Finally, I was able to save up enough money from my job as a secretary to buy an apartment. Miriam moved out of the dormitory to live with me.

It was a good period. We bought little pieces of furniture. We entertained men. We felt very grown-up, very independent.

I told her whoever gets married first gets to keep the apartment.

A number of lesser "stars" of the Nazi era who had fled Europe to other Latin American countries such as Paraguay, Bolivia, Chile, and Brazil were now considerably more at ease than they had been at the start of their exile. And those who, like Mengele, had remained in Argentina in spite of the uncertainties were relieved to find the new government was also willing to provide them with protection. Throughout Latin America, war criminals enjoyed the tacit protection of the local tyrants, who liked to model themselves after the leaders of the Third Reich. The more prominent Nazis hobnobbed with the dictators and made themselves useful by serving as consultants to the military and the secret police. The Latin strongmen were delighted to have such specialists in the arts of repression and brute force available to them.

One of the most popular Nazi refuges was Paraguay. Its leader, General Alfredo Stroessner, structured his army along German lines, infusing his Guarani Indian soldiers with Teutonic values of discipline and ruthlessness. The Paraguayan soldiers were trained in Nazi techniques by men who were unquestionably experts.

MOSHE OFFER:

When I was drafted into the army, I was forced to remember all I had tried to forget—Auschwitz, the war, the death of Tibi. All my past caught up with me.

They tried to teach me how to fight, but I was terrified of bullets. I was constantly anxious. I developed severe emotional problems.

I didn't get along with other soldiers. I was always by myself, and the more time I spent alone, the more I thought about the war.

Finally, I suffered a nervous breakdown.

In many ways, 1956 was a watershed for Mengele, the year he "came out." He was tired of posing as Helmut Gregor; he wanted to be Dr. Mengele again. Sometime that year, he had his name legally changed from Helmut Gregor to Josef Mengele. The name change required him to admit that he had been living under a false alias. He marched into the German embassy and requested a new passport, to be made out under his own name. Years later, when the hunt for Mengele would begin in earnest, his 1956 passport photograph would be the only authentic clue the world had as to how the Angel of Death of Auschwitz really looked.

In Germany, Rolf received pictures of his "uncle" looking more handsome and distinguished than ever. The young boy was filled with admiration for his faraway relative, and longed to meet him.

OLGA GROSSMAN:

Our mother suffered very much during the years we were away in England—she missed us terribly. We lived apart for nearly six years. Dr. Shoenfeld would send messengers to Israel, who gave her news about how we were doing and assured her we were fine. But that obviously wasn't enough.

We stayed on, and kept hoping that next year we would be in Israel with our mother. Year after year passed, and we waited and we hoped.

❧

VERA GROSSMAN:

One day, we heard our mother was very sick. She thought she was going to die and was asking for us.

Our stepfather's business had collapsed, and Mother was very afraid she couldn't feed the children. She had four babies—two born in Czechoslovakia and two in Israel.

She wanted to see us—she said her last wish before she died was to have her twins by her side.

❧

OLGA GROSSMAN:

We were told we were to be sent home to Israel. We were so excited. Israel meant love, caring—it meant being reunited with our mother again.

His identity and sense of respectability regained, Mengele also longed for a reunion with his son in Germany. Rolf was now a handsome boy of twelve or thirteen, still unaware of the fact that he had a living father. The Mengele family, which had stood by Josef through these difficult years, decided to help bring them together. It was the trusted Sedlmeier who played the key role in working out the logistics. He personally went to Freiburg to escort the excited little boy to Switzerland for what was to be a special winter holiday.

In Engelberg, a small, exquisite ski resort high in the Swiss Alps, Rolf met his cousin Karl Heinz and his beautiful aunt Martha, the widow of Josef's brother, Karl. Standing next to the two was "Uncle Fritz," looking even more dashing in person than in his photographs.

The hotel that was selected for this momentous occasion was the best the Mengele family could afford—an elegant Old World establishment where Queen Victoria herself had once stayed. For the first time ever, Rolf received pocket money—compliments of Uncle Fritz. As lavish meals topped off with spectacular desserts were brought to his uncle's room by deferential servants, Mengele's child was absolutely beside himself with joy, a grown-up Rolf would later admit.

VERA GROSSMAN:

When we got home, we were shocked to find there was nothing to eat —our family was extremely poor.

I remember the Friday night dinner after we arrived. Mother opened a can of sardines—each of us got one half of a sardine. And then there was some sort of thin bean soup. This was our first Sabbath meal at home.

It was a magical holiday. The best times were early in the morning and late at night, when Rolf got to crawl into bed with his "uncle" and was regaled with war stories about the Russian front and more tales of gauchos in South America. For the brief period they were together, Mengele showed himself to be a wonderful companion. Indoors, Mengele was a spellbinding storyteller. And outside, he showed himself in excellent form on the slopes. Rolf, who had not been told the truth and had no idea the wonderful visitor from South America was his father, was bursting with pride. As might be expected of a young, fatherless boy, he was utterly beguiled by his handsome uncle Fritz.

Those few days at Engelberg bought the old Nazi his fiercest, most devoted ally. Like the twins, Rolf never forgot Mengele. But Rolf's memories were of laughter on the ski runs, of chilly mornings when he padded over to Mengele's room and snuggled next to him under the great eiderdown quilt. Mengele was warm, playful, and loving toward him throughout the brief holiday.

Even as an adult, after he was made aware of charges that "Uncle Fritz," his father, had been a cold-blooded killer, Rolf Mengele could never get over that admiration bordering on hero worship. In interviews with the press, he sought to conceal such feelings. He suggested he accepted the veracity of the stories about Dr. Mengele. But it was clear he could not help loving the man who had haunted his childhood. Rolf always saw his father through the prism of his old letters, the brightly colored stamps, and the photographs that had enlivened his Black Forest childhood.

Years later, as a lawyer with his own private practice, Rolf would cling to legal principles that would help his father's "case." In law school, he was taught that a man was innocent until proven guilty. And guilt could not be determined until a trial was held. This circular thinking provided a convenient rationale for him to defend his father. For, of course, there had never been a trial of Josef Mengele—witnesses had never been able to submit their painful testimony before a court of law. When asked whether he felt his father should have been punished for his deeds, Rolf Mengele always answered with perfect equanimity, "If he was guilty." What lay underneath the impeccable legal stance had nothing to do with the law: It was the love a son felt for a father, even a father accused of heinous war crimes.

Rolf later found himself yearning for that brief childhood holiday. "It was my best vacation ever," he said nearly twenty years later. "I had . . . other uncles, but none of them were as nice as this one," he told journalists. When it was revealed in 1985 that Josef Mengele was dead, it was Rolf, alone among the Mengeles, who told the world the "true story" of his father's life.

That story was not about sadism and bestiality. It was not a portrait of a pathological killer. Rather, Rolf released old family pictures that showed Mengele appearing kindly and benevolent, his arms around his son, on the slopes with Karl Heinz, and embracing Aunt Martha. In other, older photographs, young Beppo cycled merrily about the Bavarian countryside in knickers and an overcoat. There was nothing in the mountain of innocuous letters, sentimental notes, pictures, and poems that Rolf released that shed any light on his father's crimes, nothing that mentioned the death camp. Indeed, the son paid little attention to his father's crimes at Auschwitz, acknowledging only reports that they had occurred. Rolf did say his father had sworn to him he had never killed anyone "personally."

If SS doctor Josef Mengele had needed a lawyer to defend him after he died, he could not have retained anyone more quietly eloquent, more gently persuasive than his own son. For no amount of Mengele money could buy the motivation his love for "Uncle Fritz" had given Rolf.

The Switzerland vacation over, father and son went their own separate ways, the father to his life in Buenos Aires, the child to school in Freiburg. After Mengele's return to Argentina, many more letters flowed back and forth between him and Rolf. The young boy continued to believe he was corresponding with a beloved uncle.

In the next few years, Mengele traveled to neighboring countries such as Paraguay for his business. He spent a lot of time in Asunción, where his good friend Rudel lived. He met other German expatriates through the Nazi pilot, including Werner Jung, who claimed to be head of the Nazi party in Paraguay, and Alejandro von Eckstein, a captain in the Paraguayan Army and a close friend of Stroessner.

Once again, Mengele displayed the confident air of his youth. At the offices of Fadro Farm, he was pleasant and charming to the employees and exceedingly well dressed. Colleagues recall that he liked to whistle opera tunes as he worked.

MOSHE OFFER:

One day, I started screaming, "Heil Hitler! Heil Hitler!" *I was immediately placed in a psychiatric hospital in Haifa called Blumenthal.*

The Blumenthal Hospital was very special—it specialized in treating Holocaust survivors.

I spent nearly two years at Blumenthal. I was in a terrible condition. They gave me shock treatments. I was forced to take a lot of medications. At one point, I was taking fifteen pills a day.

But everyone was very kind to me over there. They helped me as much as they could.

When I left Blumenthal, I felt very much alone. I would walk around the streets by myself. I was like a bum.

And no one in Germany seemed interested in arresting Dr. Mengele—or any other Nazi war criminal. Other countries like the United States were allowing former Nazis to settle and blend into mainstream culture. Some of the Germans the United States recruited for its budding space program, such as Walter Rudolf, had run factories during the war manned by slave laborers. But their pasts were swept aside as they built American rockets that could compete against the Russian-designed space technology, and they became honored, valued citizens of this country.

Mengele's prospects, seemingly so glum only a few years back, had brightened considerably. In many ways, the exile years in Argentina had proven to be idyllic. Dr. Mengele once again enjoyed friends, money, status, and considerable influence.

8

THE ANGEL RETREATS

VERA GROSSMAN:

Times were very hard for us in Israel. Our family had no money. We moved to a horrible slum in the Arab section of Haifa. It was the only apartment our parents could afford.

The window of my room overlooked a courtyard. I would watch as drug addicts injected themselves. It was terrifying.

Our parents had four babies to support, but there was nothing to eat in the house. It was so different from the life we had known in England. Yet in England we had also suffered, because we missed our mother.

Olga and I were forced to abandon our studies and go to work. We were only fifteen years old. Our entire schooling consisted of the five years we'd spent in England.

I had dreamed of studying drawing. Instead, I had to take whatever job I could find to earn some money.

I took a hundred different odd jobs. I did anything and everything to earn a bit of money. I worked in an ice-cream factory. I became a dressmaker. Olga got a job as a hairdresser. Whatever we earned, we gave to our mother—every last penny. We never bought anything for ourselves.

I remember going to stores and smelling the food because I was hungry, yet buying nothing. I would take the money and tell Mother, "Don't worry about me, I don't need anything."

One day, a matchmaker introduced me to an American boy who was looking for a bride. He was a very, very rich young man, and he was prepared to take me to the United States.

He liked me, and tried to persuade me to marry him. He offered me a good life in America. But I refused.

When he persisted, I finally told him, "Look. I have been through hell living apart from my mother. Now that I am with her, I am not going to leave her to go to America with you."

Yet the situation at home kept getting worse. Olga was very sick. And no one could figure out what was wrong with her. She kept fainting.

❧

OLGA GROSSMAN:

On Yom Kippur, I passed out at the synagogue. In the middle of services, I fell to the ground. My mother panicked and asked, "Could someone please bring some water?"

A soldier in uniform came out of the crowd, and brought me a glass of water. His name was Rafael, and he was the only person in the entire congregation who bothered to come to my aid.

I saw him like a dream when I woke up. He was so handsome.

We got to know each other when I went to thank him. I was very shy. I could hardly look at a boy. But I realized he was everything for me. As far as I was concerned, he was my guardian angel—I called him my angel Rafael.

It turned out we had similar backgrounds. He, too, had been at Auschwitz. He was the only one in his entire family to have come out alive. He lost his parents, his two brothers, and a sister.

There was an instant kinship between us, but he didn't propose immediately. I was only seventeen; he was ten years older. He was a professional soldier. I liked that. He was very masculine, very tough. It gave me courage, seeing him in uniform.

When he asked me to marry him, my parents were worried because

I was so young. He promised them he would wait—until I turned eighteen. We were married the following year.

The ski trip to Switzerland gave Mengele an opportunity to get to know not only his son but also the woman who would later become his bride. Martha Mengele, the tall, attractive widow of his brother Karl, had got along famously with Josef. Martha moved to Argentina with her twelve-year-old son, Karl Heinz, in the fall of 1956. They settled in Mengele's impressive villa in Olivos, near Perón's former residence. Photographs they sent back to the family in Günzburg showed a dashing couple. At last, Mengele had the "family unit" he had craved—husband, wife, and child together again.

On the surface, Martha and Josef seemed a perfect match. Like Mengele, Martha was a triumph of style over substance. During her Günzburg youth, she had been an object of admiration because of her lithe figure and beautiful face. At thirty-five, she took pains with her appearance, and looked, in the words of an old acquaintance, "like a fashion model." The couple was urged to marry by relatives, particularly Karl Sr. It is unclear how well Martha knew her brother-in-law before she gave in to Karl Sr.'s pressure. Certainly, by marrying Josef, she solved two of his most pressing concerns: retaining the firm in Mengele hands and caring for Beppo in South America. The Mengeles had been worried that popular Martha would remarry outside the family.

JUDITH YAGUDAH:

As a young girl, my mother was so overprotective, she didn't even want me to go out on dates. I could not make any friends. If a boy invited me to the cinema, she would say to me, "Why should you go out to the movies with him tonight, and leave me here alone?"

I invariably turned the boy down.

Mother was always talking about Ruthie. She would tell me, "If I had the other twin, it would be much easier. But I only have you. My love is only for you."

I was very close to my mother—too close. She would not let me lead a normal life. I was very much under her control. Our relationship was not healthy—especially her attachment to me.

The native Israeli girls, the "Sabras," were so free, so sure of themselves. They had lots of friends. They went out with boys, alone and in groups. But I stayed home with my mother.

One day, I was invited to a party by someone who was also from Eastern Europe. This time, I accepted the invitation.

That's how I met my husband. He was from Yugoslavia, and also a concentration-camp survivor.

He had been at Bergen-Belsen. He lost both his mother and father when he was thirteen. The Nazis murdered them in front of him. He watched them being shot, and he remembered.

We began going out. Mother didn't like our relationship, of course, but I continued to see him. We went out for three years, and then he proposed. He was the only man I ever dated.

Mother didn't want me to marry him. She'd point out how poor we both were. "You have nothing and he has nothing," she'd say. "What will become of you?" She thought I should stay with her until I found a better match.

I decided to marry him anyway.

In 1958, Martha and Josef decided to get married and flew to Uruguay for the ceremony. News of the nuptials caused a stir in Günzburg. Josef's boyhood friends were astonished that he had married his late brother's wife. Those who knew Martha were taken aback—not by her decision to marry yet another Mengele, but to live in some South American backwater. In Günzburg, Martha was seen as a playgirl, and something of a gold digger. Years earlier, she had become pregnant by Karl Mengele, Jr., while she was married to one of his best friends, a local businessman named Wilhelm Ensmann. She promptly divorced Ensmann and married Karl. The baby, Karl Heinz, was the object of a paternity suit, but a court ruled he was Karl's child. Martha's strong predilection for the good life was well-known. It was hard to imagine her doing without the comforts of Europe.

VERA GROSSMAN:

My husband, Shmuel, came from a family that was even poorer than mine. I met him through another matchmaker.

He lived with his parents in an Arab village. His house was so run down that when I spent the night there, I kept thinking the roof would cave in.

But he was a very good-looking boy, very nice and understanding. He was six years older than I, and worked as a pipe-fitter. We liked each other immediately, and he proposed.

But our engagement wasn't official until my stepfather gave his blessing. When my stepfather met Shmuel's parents, they realized they had known each other in Poland before the war. My stepfather said they were a very good family, and he approved the union.

The wedding took place only three months after we met. His family had to sell a cow to raise money for the ceremony.

We rented the cheapest apartment we could find in Haifa. It was a little shack perched on top of the roof of a five-story building in the Arab section. There was a little room where you could fit a bed and nothing else—not even a dresser or a table. The bathroom was practically in the kitchen. It didn't even have a closet.

On my wedding day, I went over to the apartment to try to fix it up. I found a space in the wall where I decided to build a makeshift closet. I took a hammer and broke down the wall all by myself, and constructed a small closet.

I marched down the aisle with blisters on my hands.

We couldn't afford a honeymoon, so we moved into our apartment that same day. But first, I had to return my bridal gown, which I had borrowed from a friend.

After the wedding, I went by foot to return it—I walked two or three kilometers because I couldn't afford a taxi. Then I walked several kilometers more back to our new apartment.

❧

MOSHE OFFER:

I met my first wife shortly after leaving the Blumenthal Hospital. I was seventeen, she was eighteen.

We fell madly in love, and decided to get married immediately.

She became pregnant shortly after our wedding. I was very happy for the first time since the war. At last, someone was taking care of me. I had no idea she herself was not well.

I learned, too late, that she had a heart condition. She should never have gotten pregnant. She died in childbirth. She left behind a healthy baby girl, our daughter.

I was devastated. I had loved her so much. She had saved my life.

After she died, I didn't know what to do. Finally, I showed up at the door of Blumenthal, carrying our baby in my arms.

Josef and Martha's bliss was short-lived. Within months of their marriage, he was picked up by the local police. When he had first arrived in Argentina eight years before, he had treated the sick daughter of his roommate at the pension in Vicente Lopez. The authorities were investigating him for practicing medicine without a license. Mengele was also implicated in a ring of doctors who were illegally performing abortions. He was questioned by police and held for a few days, but the charges were apparently dropped.

Mengele's life was marred by yet more drama when his father died in 1959. The news was a terrible blow to Josef, who lost his most loyal supporter in the Mengele family. Karl Sr. had never stopped loving his son, in spite of the growing evidence of Beppo's appalling crimes during the war.

The death of Günzburg's leading citizen—and its largest employer—prompted a funeral procession that stretched across town. Hundreds of people came to pay their respects to the man who had done his best to make the Mengele name honorable again, and Josef is believed to have made a quick, furtive trip to attend his father's funeral. Although the reports are contradictory, a couple that was friendly with Mengele at the time told German prosecutors that he was indeed in Günzburg to pay his final respects. A large wreath appeared mysteriously on the grave of Karl Mengele, bearing the inscription, "Greetings from someone far away."

The death of Karl Sr. came at a time when Mengele's life was again stirred by turmoil. In addition to his problems with the Argentine police, he was once again a wanted man. That same year, 1959, the German government made an effort to locate the Auschwitz doctor. It was spurred by a new breed of Nazi-hunters who were trying to track down old Nazi war criminals: concentration-camp survivors determined to avenge the deaths of loved ones by going after those who had escaped punishment.

In Vienna, an Auschwitz survivor named Hermann Langbein had vivid memories of the Angel of Death standing at the head of the selection line. As secretary-general of the International Auschwitz Committee, Langbein corresponded with other Holocaust victims and tried to discover the whereabouts of major war criminals. He had always been particularly bothered by the fact that Mengele had eluded the courts—and the gallows. Langbein's initial attempts to find out where Mengele had gone after the war proved fruitless. Ironically, it was Mengele's divorce from Irene that enabled Langbein to track down the death-camp doctor to his South American retreat.

Irene and Josef's divorce papers, which had been filed in a Freiburg court in 1954, had conveniently included a document listing Mengele's address as Buenos Aires. Clearly, neither Josef nor his family thought it risky anymore to list such revealing information. As it turned out, this was one of the very few mistakes the family made: By uncovering these documents, Langbein ascertained that Mengele was alive and well, and residing in Argentina under his own name.

LEAH STERN:

Some people tried to persuade my husband-to-be not to marry me because of my past as a Mengele guinea pig. The two of us had met on the ship to Israel, and were friendly before we were romantically involved.

His friends asked him why he was taking a chance settling down with a victim of Mengele's experiments. They pointed out I might have trouble having children.

But he loved me very much. He was determined to marry me whether I could have children or not.

❧

MENASHE LORINCZI:

My future in-laws were very worried about my past as an Auschwitz twin; they wondered whether I was "healthy."

When I first met my wife, Yaffa, I thought she was so beautiful, I couldn't take my eyes off of her. I wanted to marry her then and there.

My family and I had recently moved to Netanya. She was a native Israeli who had lived in the city all her life. She noticed me because I was the "new boy in town."

I learned Yaffa was involved with another guy, but that didn't put me off in the least. I had a friend of mine ask her point-blank if she was serious about the man. She told him she wasn't. The next day, I asked her out on a date, along with another couple.

At first, we double-dated a lot, or went out in a group. We'd go to the beach, to the theater, to the movies. Although I was in love with her, Yaffa wasn't sure she wanted to marry me. She came from a very religious background.

As I began turning up at her house more and more often, her parents asked her who I was, and whether my intentions were honorable.

Even after I proposed—and she accepted—new problems cropped up.

Yaffa's parents were worried because I was very skinny, and thought I might be sick because of my time at Auschwitz. I had not talked to them very much about my experiences in the death camp. At that time, survivors kept their mouths shut about the war.

Some people told my wife's parents I had been at Auschwitz. They were advised to think twice about letting their daughter marry me—because of my past as a Mengele twin.

Langbein set about amassing a file on Mengele and his crimes at Auschwitz, in the hope of having him extradited and tried. But when he presented his dossier to the West German government, he found the bureaucrats reluctant to reopen the case. West Germany had been out of the Nazi-hunting business for years, and not even the possibility of catching the infamous Dr. Mengele could goad them to action. Langbein persisted, until he finally got a prosecutor in Freiburg interested in the case. Germany issued its first arrest warrant for Mengele on June 7, 1959. The German Foreign Ministry was forced to seek his extradition from Argentina.

This confluence of events persuaded Mengele he could no longer safely remain in Argentina, and that he needed to find a new home for himself, Martha, and Karl Heinz. He settled in Paraguay where his friends in the German community gladly offered him refuge. The future seemed perilous and uncertain—an unpleasant reminder of former days. The Argentine idyll was over. He was on the run once again.

PETER SOMOGYI:

One day, our father said we should leave Israel. He had never been able to open a successful business there, as he had in Eastern Europe. He felt there were better prospects elsewhere.

But I didn't want to leave Israel. It took a lot of convincing on the part of my father to get me to go.

My twin brother and I went to London. We applied to emigrate to any country that would take all of us, where we could have a new life. Canada was our ideal. Our father had decided we would have better opportunities in Canada. He would join us after we were settled somewhere.

Meanwhile, we needed to work to support ourselves in England. It really helped that I had a trade. I got a work permit, and I quickly got a job as an automobile mechanic.

My brother had a harder time. He took odd jobs to earn money while he went to school. At one point, he was a garbage collector at Marks and Spencer.

Both of us hoped and prayed we would get papers to emigrate to Canada.

❧

EVA MOZES:

I knew my future husband only ten days when I decided to leave Israel and settle down with him in America.

Mickey was also a Holocaust survivor. Like me, he had lost both his parents in the concentration camps. Although he was originally from Riga, in the Soviet Union, he had settled in Indiana.

I knew his brother, who lived near me in Israel, very well. His family had plotted the match. They had wanted me to meet him, and had prepared me for his visit months before he got to Israel.

When we met, we found we couldn't even speak the other's language. I knew very little English; he didn't know Hebrew. We talked with two dictionaries. But I toured Israel with him, and we managed to have fun.

During the ten days, he pressured me to agree to marry him. I told him, "Return to the U.S. and we'll correspond—I can't make a decision that fast." But he said no; if I wanted the relationship to continue, we had to get engaged.

My sister, Miriam, was married and had a baby. All my friends were married. I was twenty-six and still single, which was practically unheard of in those days. My aunt kept pressuring me to say yes. "Don't be an old maid," she'd tell me. "Get a divorce, but get married."

I was still heartbroken over a recent affair I'd had. Whatever I did, I could not forget the man. What made it worse was that he continued to see me, even though he was married now. He would pop in at any moment.

It was a great love, but I knew it had no future. The thought of leaving Israel was very appealing. There was adventure in traveling to a new place. Going to America seemed wonderful.

I told Mickey, "Okay, I'm going," and left Israel on a plane bound for New York.

❧

LEA LORINCZI:

In 1959, my husband convinced me to leave Israel and go to America with him. My husband had always wanted to live in the United States—that was his big dream. But he didn't tell me that until after we were married.

At first, I fought him. I didn't want to leave my father and my twin brother. But he insisted. "Let us try it," he told me. "If it doesn't work out, we can always come back to Israel."

❧

ZVI THE SAILOR:

When I came to Israel, I joined the navy. That was how I came to realize I loved the sea, and decided to become a sailor.

At first, I saw sailing as a means to an end. I knew I had an aunt in America, and I wanted to visit her. Afterwards, I continued to sail because I wanted to run away from Israel, and perhaps from myself. And then, I kept sailing because I was used to it and it was an easy life.

When I started sailing, I made plenty of money. The more I made, the more I spent. I was twenty years old, and I thought I was very old.

I led a good life—the life of a vagabond. There were women in every port. We were always getting drunk.

For those who live on land, buying a chair, a table, with the money they earn is a mark of success.

But as a sailor, I spent money simply on having a good time. And that was okay, because you never knew if you would return from the next voyage.

Mengele's precise movements and whereabouts in the period between 1958 and 1960 are difficult to ascertain: the findings of scholars, journalists, investigators, and intelligence specialists concerning this period are often contradictory. What is known is that as the search for Mengele intensified, he retreated from Argentina to Paraguay, from Paraguay to Brazil, making trips between the three countries before finally finding a safe haven in Paraguay in 1959. The regime of Paraguayan strongman Alfredo Stroessner was even more accommodating to Nazis than Argentina's had been under Perón. When Mengele applied for Paraguayan citizenship in the fall of 1959, two good friends swore to his worthy character: Werner Jung, head of the local Nazi party, and Captain Alejandro von Eckstein, the White Russian known for his fascist views and who had first been introduced to Mengele by their mutual friend Hans Ulrich Rudel. Both men claimed that Mengele had been in Paraguay for the five-year period requisite for citizenship. Von Eckstein, who had fought in the Chaco Wars under then-Captain Stroessner, was able to draw on his connections with the dictator, enabling Mengele to obtain a Paraguayan ID—under his own name—and become naturalized as a citizen.

Mengele continued to return to Argentina even after he obtained Paraguayan citizenship. He still had substantial interests back in Buenos Aires, including his villa in Olivos and his share in the Fadro pharmaceuticals company, and had to make frequent trips to liquidate his various holdings. He even went back to work at Fadro for a period. His former co-workers thought he seemed rather glum.

Although in the course of these sojourns there were occasions for the Argentine and German governments to nab him, Mengele remained free. The Argentines who had provided Mengele a safe haven all these years were certainly not anxious to extradite him. As for the German diplomats in Buenos Aires and Asunción who were handling Bonn's requests for information on Mengele, they, too, cast a cold eye on the initiatives to capture the Auschwitz doctor.

Newly declassified files of the U.S. State Department provide a fascinating glimpse of the minuet the German and Argentine governments performed to help out the Angel of Death. According to cables sent by the U.S. embassy back to Washington in June 1959 (while Mengele was shuttling back and forth between Buenos Aires and Asunción), the German government asked Argentina to begin proceedings to allow Mengele to be extradited. The Argentines coolly replied that their own inquiry had revealed "no record of the subject's entry into this country." This was despite the fact that Mengele was listed in the Buenos Aires telephone book under his wife's name. With remarkable gall, Argentine officials pressed Germany for "additional information" to support their criminal allegations against Mengele.

The Germans didn't respond until six months later, when they sent Mengele's address to Argentina. Still, Argentine officials stalled. Possibly the bureaucrats felt that if they delayed long enough, the extradition request would fall by the wayside. More practically, sympathetic officials may have wanted to give Mengele additional time to arrange his departure. After all, the death-camp-doctor-turned-executive had a great deal of business to attend to in their country.

When they did respond, the Argentines argued there was no formal extradition treaty between their country and Germany, and hence no legal mechanism they could use to accede to the Germans' request. Instead, Buenos Aires officials said the case should be submitted to their solicitor general "for a recommendation." This wasn't done until June 1960—one full year after Germany's initial request for extradition. "But by then, Mengele, who had finally been located in this country (Argentina), had disappeared," read one of the cables from the U.S. embassy. The various delays had given the war criminal more than ample time to plan for his future and to take care of any outstanding business interests. According to an Israeli diplomat who was then assigned to Buenos Aires, Mengele even made a nice profit on the sale of his luxurious villa in Olivos.

JUDITH YAGUDAH:

For the first five years I was married, my mother, my husband, and I lived together in a cramped apartment in Haifa. It was awful, the three of us squeezed under one roof. But Mother wouldn't have it any other way.

We could have afforded another apartment for her, but she refused. She was afraid to live alone.

What made it even worse was the fact that I was working to support us, while my husband was studying at the university. My mother thought it was shocking that a woman would have to go to work. She made it very difficult for me.

I was in conflict the whole time. I was torn between my mother and husband. On the one hand, I wanted to be free, and lead my own life. On the other hand, I had been brought up that one should feel responsible toward one's parents. My poor husband had no choice but to accept the situation.

❧

LEA LORINCZI:

We led a hard life at first in the United States. It was very difficult to establish ourselves financially.

Even though my husband didn't want me to work, I decided to get a job. Since I couldn't speak English, I was not able to be a nurse. Instead, I was hired at a sweater factory. I had to work hard to make ends meet.

I felt very lost. I had left my family, and my new life was very trying.

It took five years before we could afford to go home to Israel.

Hidden away in Paraguay, at the home of his friend Alban Krugg, Mengele was planning his next move. During this period, Martha kept hoping they would resume their life together: she thought Buenos Aires was still safe for all of them. But Mengele, cautious as ever, had no such illusions. He planned to wait until his pursuers either got tired or despaired of being able to find him. His instincts to lay low proved to be absolutely right. By now, the Angel of Death was an expert on avoiding capture.

Even as Mengele retreated out of sight, a small crack team of Israeli agents prepared to nab another ranking Nazi, Adolf Eichmann, the engineer of the Final Solution. They apparently hoped to capture Mengele at the same time. Eichmann was quietly whisked off the streets of Buenos Aires one evening in May 1960 as he returned from work. The agents persuaded Eichmann to tell them Mengele's address, but by then it was too late. Mengele was safely ensconced in his Paraguayan retreat.

On May 11, 1960, the capture of Eichmann was revealed to an incredulous world. Eichmann had been living under the pseudonym of Ricardo Klement. Each day, the former Gestapo colonel took a bus to his dreary job. Unlike Mengele, Eichmann had no family money to invest in a business of his own or to vacation within Swiss ski resorts. The man in charge of rounding up all of Europe's Jews for the death camps was now nothing but a lowly clerk eking out a living to support a wife and four children. Instead of residing in a fine villa such as Mengele's, his family was squeezed into a little house that had neither running water nor electricity.

Mengele, for whom such things had always mattered, had not really socialized with Eichmann, presumably because his social status was too low. But they did share contacts in the *Kameradenwerk*, the secret organization that gave aid to former Nazis. Wilhelm Sassens, the journalist for *Der Weg* who had helped Mengele in his early days, had also befriended Eichmann.

After the kidnapping, Martha decided she had had enough South American adventures and returned to Europe with little Karl Heinz. It was clear to her that she and Josef would never be able to lead a normal life together, and she had no interest in following him to his next place of refuge: Brazil. Life in the cultured, cosmopolitan Argentine capital had been tolerable for Martha; Paraguay and Brazil would be impossible. Martha's complaints about her rough life in the backwaters of South America became the talk of Günzburg. Residents of the sleepy town still chuckle when they remember Martha whining about her ordeal as the bride of Josef Mengele: Why, she had even been forced to do housework. Somehow, the image of sleek Martha mopping floors and scrubbing pots and pans in the wilds of Paraguay, Argentina, and Brazil engendered more derision than sympathy.

VERA GROSSMAN:

After we were married, my husband and I continued to support our families—even though we had barely enough money to live ourselves. I suffered, but I never complained. I told myself, As long as I am alive, it's okay. My optimism helped me to overcome the hard times. I got through the hard times because of my spirit. I would tell my husband, "I don't mind being poor." With hope in your heart, you feel optimistic, and you overcome every hardship.

In his Paraguayan hideout, Mengele wrote about the Eichmann capture in his journals, saying he now had to be even more careful than before. The good old days when he could saunter into the German embassy and coolly request an ID under his own name, then jet off for a European holiday, were gone forever. In the fall of 1960, Mengele moved to Brazil. His friend Rudel had set him up with a man who could help him, an unrepentant Austrian Nazi named Wolfgang Gerhard. Gerhard worked out a safe living arrangement for the Auschwitz doctor. Mengele went into hiding again, and remained out of the public eye for the rest of his life.

The Eichmann capture seemed to augur a new era where Nazi-hunting would become a more aggressive pursuit. Men like Simon Wiesenthal and Langbein, both working out of Vienna, sought to pressure the German government, eager to forget the war and now securely launched into postwar reconstruction, to do more. Their persistence, combined with the anti-Nazi sentiments fueled by the Eichmann kidnap, jolted Germany out of its lethargy. In the fifteen years since the end of the war, the Germans had appeared reluctant to punish the perpetrators of the Holocaust. The Central Agency for Investigation of War Crimes, which had been assembled in 1958, had in two years accomplished very little.

But months after Eichmann's capture, the Germans revealed some dazzling Nazi-hunting skills of their own. Richard Baer, the commandant of Auschwitz after Rudolf Hess, was arrested, as were members of the Eichmann "team." There was a new element at work that added impetus to Germany's initiative to hunt down Nazis. The Eichmann capture and trial had stirred up deep feelings of guilt and shame in many Germans. Gideon Hausner, the Israeli prosecutor, received many moving apologies from individuals throughout Germany in the course of the trial. One young German wrote Hausner that he wanted to "atone" for what his elders had done "against humanity." He offered to work in Israel, and indicated his group of friends were ready to do the same. Their only question was, "Will you take us?" One German family wrote that they had been inspired by the revelations of Nazi atrocities to go see Dachau. Deeply moved, the family emerged from their tour of the death camp convinced that "whatever punishment was meted out to Eichmann was not enough."

MENASHE LORINCZI:

Until the Eichmann trial, survivors never talked about the Holocaust because nobody believed them. When I was in the army, I tried to tell people what had happened to me at Auschwitz under Dr. Mengele. "Are you crazy?" the other soldiers said to me.

❧

HEDVAH AND LEAH STERN:

The Eichmann trial destroyed us inside. Everything came back—the dead bodies they piled up in front of our barracks, day and night, day and night, day and night. We kept thinking about that throughout the trial.

Until then, no one spoke about what happened in the war. We would tell friends our stories, and nobody believed us—nobody.

❧

ALEX DEKEL:

I went to see Eichmann, face-to-face, in his cell the day before he was executed. I asked him if he remembered me. He said no. Then I rolled up my sleeve and showed him the tattooed number on my arm. He turned away.

❧

MIRIAM MOZES:

After the war, it was something shameful to admit you were a Holocaust victim. Nobody wanted to talk about it.

I was lucky. I was able to confide in my husband from the start. He was a good listener. I told him the whole story of what had happened to me at Auschwitz as a Mengele twin. He was a Sabra, and he had not lived through the war, but he was very interested—even in those years when nobody cared. He wanted to know all about the Holocaust. He sympathized with what I had gone through.

He would tell me, "How could a child without a mother or father return to the world, go to school, get married and lead the life of a normal person?"

He made me feel like a heroine.

But in those years, I didn't feel especially heroic. No one wanted to listen to my story. I had aunts and uncles, and several cousins. They never asked me what had happened to me at Auschwitz. They never wanted to know. And my feeling was that nobody wanted to know, except my husband—until Eichmann.

During the Eichmann trial, this changed dramatically. Even though there was no television, only radio and newspapers, we all followed it. Some survivors were so upset by the revelations, they committed suicide after the trial.

The Germans refrained from requesting Eichmann's extradition from Israel, presumably so as not to embarrass the Israelis or themselves. Germany had abolished capital punishment, and there was the disturbing possibility that even if they tried Eichmann, he would receive a lenient sentence.

This, indeed, had been the outcome in the trials of most of the Nazis Germany had caught since the end of the war. Eichmann's legal adviser had been sentenced to only five years' hard labor. Otto Bradfish, who killed fifteen thousand Jews as part of the *Einsatzgruppen*, the mobile killing units in the East, got ten years' hard labor. And Josef Lechthaler, who had slaughtered Jews in Russia, got only three-and-a-half years behind bars.

But however minimal Germany's reckoning with its past, at least arrests were made. Germany even managed to apprehend a few high-ranking Nazis still at large. Karl Wolff, for example, who had served on Heinrich Himmler's personal staff, was nabbed and tried. He was quoted as saying how he had greeted "with particular joy . . . the news that for two weeks now, a train has been carrying every day members of the Chosen People from Warsaw to Treblinka."

The seizure of Eichmann sent Nazis who were hiding around the world deeper into the underground. Like Eichmann—and Mengele—many of Hitler's henchmen had sought refuge in South America, posing as simple citizens. They were in close touch with one another by way of what Hausner called a "grapevine that spanned the South American continent and reached into the remotest parts of the pampas and jungle." From then on, every one of them would be haunted by the fear that a team of Israeli agents was on his tail, ready to pounce on him.

Around the time of the Eichmann kidnapping and trial, Rolf Mengele was told by his stepfather, Hackenjos, that "Uncle Fritz," the man he had corresponded with and loved from afar, was really his father. Interviews Rolf has given suggest he was deeply shaken by the revelation, coming as it did on the heels of newspaper stories depicting Mengele as a monster, and a perverted, ruthless war criminal. German newspapers were replete with sensational accounts of the Auschwitz Angel of Death, and young Rolf, fifteen at the time, was teased about his last name.

But as Rolf later told journalists, his family insisted to him the media accounts were wrong. He was told he should be proud to have Josef Mengele as a father. The Mengeles told Rolf that Josef was a good and brilliant man who spoke Greek and Latin fluently and had earned several advanced degrees. Rolf's own cousin, Karl Heinz, who had lived with Mengele in South America, confirmed what the rest of the family was saying. Mengele was clever and caring. He had been a good father to him, strict, but extremely fair, Karl Heinz insisted.

It was difficult for the teenager to try to reconcile the two images of Dr. Mengele—that provided by his family and the one offered by the rest of the world. Anguished by the revelations, he began failing at school. His teachers attributed this to the trauma of being the son of Josef Mengele. The tenor of his correspondence with the man he had known all his life as Uncle Fritz changed as a result, becoming erratic and petulant, he confided in these same interviews.

OLGA GROSSMAN:

How scary it was for me to have children. All the unpleasant memories came back to me. We had watched little babies thrown into the ovens.

I was very, very worried during my pregnancies. I suffered all the time. What will come out of me? I wondered. This child, will it be normal?

I would agonize over what had been done to me by Dr. Mengele. I was sure I would have an abnormal baby—because of all the tests and experiments.

After the childbirth, when they told me I had a beautiful, healthy baby, I collapsed.

It was too good to be really true. I had a nervous breakdown and had to be institutionalized.

❧

MIRIAM MOZES:

I was very depressed after the birth of my first child. I saw other women in the hospital being visited by their mothers, and that was very painful to me. For even though I was very happy, very excited over the birth of my daughter, I felt badly [that] my mother wasn't there.

Only my husband came to see me. And Eva. My twin became like my mother. She went to buy everything for the child. She would come all the time. But I still found myself staring at the other young women, and envying the fact that their mothers would come and fuss over them.

The Eichmann case also had serious side effects that threatened the safety of Jewish communities the world over. If the capture had forced old Nazis to lay low, it also prompted their supporters to emerge. There were rabid anti-Semitic outbursts throughout Latin America, and a renewed interest in Nazi causes and ideals. The rise in pro-Nazi sentiment was an offshoot of the resentment many South American nations felt toward Israel. These countries viewed the kidnapping of

Eichmann as a violation of their sovereignty—a lawless affront to Argentina's honor. There were also angry formal protests at the United Nations. In June 1960, Argentina made a formal complaint to the UN Security Council, protesting what Israel had done.

Israeli prime minister David Ben-Gurion, who had given the Mossad, the Jewish intelligence service, the go-ahead to nab Eichmann, began to see the negative consequences of the operation. Although the trial had certainly raised the world's consciousness, he worried that there would be a backlash. In addition, Israel had been rebuked by some of its staunchest friends, including General Telford Taylor, for violating international law.

With the May 1962 execution of Eichmann, Ben-Gurion's worst fears were realized. There were anti-Semitic episodes around the world. On June 24, 1962, a young Argentine Jewish girl, Graciella Sirota, was kidnapped on her way to school. Her assailants carved a swastika on her breast and inflicted severe burns all over her body. She was told, "This is revenge. You Jews are responsible for Eichmann's death." Four days later, a male student, son of a Catholic father and a Jewish mother, was attacked by four youths who slashed swastikas in his cheeks and forehead and beat him savagely.

Secret memoranda sent by Argentine Jewry to major American Jewish organizations reflected the panic roused by the anti-Semitic incidents. There was a sense that the Nazis were capable of regrouping in South America to "recreate" the Third Reich. It was believed the Nazis were involved in a plot to install a fascist regime in Argentina, and that once in power, its first goal would be to exterminate the Jews.

Since the fall of Perón, the instability of the Argentine government had indeed helped stir up anti-Semitism in a population already predisposed against Jews. There were dozens of nationalist and reactionary groups fomenting hatred and advocating violence and other forms of retribution, an extremism that recalled some of the blackest days of Hitler's Germany. For the vulnerable Jewish community, with its many war refugees, all the ingredients were there for a terrible repeat of history. As one 1962 memorandum from a local Jewish organization noted, "Events have crystallized, particularly after the conclusion of the Eichmann trial. It has been proven to us that the Nazi minority have entrenched themselves. While today, they represent no immediate danger, they can under disturbed general conditions come to the fore . . ." The major question was what the Nazis would do to avenge the execution of Eichmann.

As Israel had to decide whether to pursue more Nazis, Ben-Gurion began to have doubts about continuing. The Mossad, however, was buoyant over the success of the Eichmann capture. Isser Harel, then head of Israeli intelligence, was especially proud. From an operational point of view, it had gone practically without a hitch. His handpicked team had entered Argentina undetected and left the country with no one guessing whom they had in tow. Although they had been prepared to kill anyone who stood in their way, there was no loss of life. With the Eichmann operation completed, the Mossad team was eager to return and mount a full-scale attempt to apprehend the elusive doctor Josef Mengele.

Harel was especially anxious to capture Mengele, whom he personally considered much more despicable than the bureaucrat Eichmann. But Ben-Gurion, increasingly fearful of the repercussions against Jews and against Israel, is believed to have counseled Harel against undertaking any more missions. Nevertheless, the feisty Mossad chief had his men discreetly continue the hunt for Mengele.

Tales of the Mossad's search for Mengele belong more to the realm of mythology than history. Details about the secret operations have been considerably embellished by the ex-agents who took part in them. There are stories of searches launched across Europe and South America and stakeouts deep in the heart of the Paraguayan jungle. Various agents who have "gone public," including Harel, Rafael Eitan, and Peter Malkin, all three of whom were on the original Eichmann kidnapping team, have given confused, exaggerated, and usually contradictory accounts of the Mossad's attempts to find Mengele. Harel himself has said that although his men did finally locate Mengele, it would have taken a small army to nab him, with only limited chances of success and the possibility of great bloodshed. Other agents have insisted the Mossad never came close to finding the Auschwitz doctor of death. One fact is undisputable: The Mossad ultimately failed in its attempts to capture Mengele.

By 1962, any impetus for continuing the hunt for Mengele was lost because of Harel's involvement in other matters. His disputes with Ben-Gurion, and resignation the following year, halted any real or imaginary plans the Mossad may have had. Relations between Harel and Ben-Gurion had been poor for years. In 1962, the final confrontation between them came over Nazis in Egypt, not South America. The Israelis learned that Egypt was starting up a rocket program with the help of Nazi scientists. Much to the embarrassment of the Mossad, information about the program surfaced only when Egyptian strongman Gamal Abdel Nasser paraded the rockets through the streets of Cairo. It was charged that Harel had focused his agency's resources on the wrong enemy.

In a show of passion, Harel offered his resignation in 1963—some say with the certainty it would be refused. Harel had built the Israeli secret service and had turned it into one of the world's leading intelligence agencies. But Ben-Gurion had apparently had enough of Harel, and resented his insubordination. The resignation was accepted.

Once again, fate had intervened to help the Angel of Death elude punishment. Isser Harel had been the person most fervently committed to the idea of pursuing ex-Nazis. He had kept a secret notebook on Mengele, recording whatever information he collected on his character traits and proclivities. When he resigned, the main threat to Mengele's survival was gone. The men who would follow Harel as head of the Mossad were also talented superspies, but none showed the same unswerving concern in hunting down Nazi war criminals. A young PLO thug was seen as a far greater threat than the aging Nazi doctor who had condemned hundreds of thousands to die with a wave of the hand.

But long after the Israelis had abandoned the chase, Mengele continued to be haunted by the fear that the Mossad was after him. Years after the execution of Eichmann, he still harbored the notion that the Israelis were plotting to kidnap him. A strange mixture of fear and arrogance persisted in Mengele even into his old age, so that he continued to believe he was the target of a manhunt. In one retreat, a remote Brazilian farm where he moved sometime in 1962, he ordered a watchtower to be built so he could observe the roads in search of the secret agents who never arrived. Farmhands would later recall how the Auschwitz doctor anxiously stalked the grounds, surrounded by packs of fierce dogs.

Mengele was determined to outwit all his enemies. If the Nazi-hunters were searching for a wealthy, dashing businessman, he would fool them by posing as a humble farm manager. They would scour the cities and jungles of Paraguay—and he would retire to an obscure Brazilian village. His pursuers would seek a mansion, protected by legions of uniformed guards, and he would be safely tucked away in an unpretentious little farm.

In the early 1960s, in his new Brazilian hideaway, Mengele was once again alone. His father was dead; his wife and stepson had left him. His beautiful Argentine villa now belonged to someone else.

Gone also was his beloved Borgward "Isabella." All the efforts he had expended in the last decade to build a new life had come to nothing.

HEDVAH AND LEAH STERN:

Our lives really began when we built our Moshav, our communal settlement. We and a group of young people decided we needed a place we could call home. There were twenty of us, boys and girls, and we had come from all over Europe. We were all Holocaust survivors, and many of us had been at Auschwitz.

We picked a site in Ashdod, near the sea. We thought we could combine agriculture and fishing. We called our commune "Nir Galim," *which means the air and the waves.*

At first, we lived in tents—then in huts. Finally, we built houses for ourselves. We worked very hard. We tilled the soil. We cooked meals in the kitchen, and washed the pots and pans. If we found a piece of wood, we were overjoyed, because we could use it to build a table. We used whatever we could find to build the Moshav.

Yet Auschwitz remained with us—everything around us reminded us of it. From our Moshav, we could see the chimneys of a factory in downtown Ashdod. We were reminded of the chimneys of Auschwitz, especially at night, when the flames poured out.

Dr. Mengele was getting on in years. His once-fine features were turning flaccid. At fifty-one, he suffered from a variety of small aches and pains that foreshadowed the infirmities of old age. The man who had condemned so many to die began to confront the notion of his own impending death. Alone and sickly on an isolated little farm, he awaited the empty days that stretched ahead, filled only with an unrelenting loneliness. Mengele was in a state of despair, anxious to leave a record of some sort behind. There was little left to hope for: with his father now dead, it was unclear how sympathetic the rest of the family would be to him.

It was in his new hideout, on a remote farm hundreds of kilometers away from São Paulo, that Mengele began to work on his memoirs. Facing a dreary and uncertain future, he found it more comforting to look to the past instead, back to the halcyon days of his privileged youth, when it seemed certain the world was to one day be his for the taking.

ZVI THE SAILOR:

I went all over the world, and I always wanted to go elsewhere—to yet another foreign city or country I hadn't seen. Cleveland, Chicago, Detroit. Nova Scotia, Newfoundland, New York, Philadelphia.

I sailed without a break. I would go from one ship to another, from one voyage to the next. I was married—and yes, it was hard on my marriage—but still, I sailed.

My wife and I bought an apartment in Ashdod, near the sea. From the window, I could see the chimneys of the factories. They made me think of the crematoriums. And so I would leave her for weeks and months at a stretch.

I sailed wherever there was water. I worked as a fisherman in the North Sea. When they opened up the Great Lakes, I was on the first ship to travel there. "What are these Great Lakes?" I wondered. I had to know.

But eventually, I found myself back at Auschwitz. I was at Auschwitz, no matter where I went.

9

BRAZILIAN HIDEAWAY

MENASHE LORINCZI:

For years after the war, the popular line was that we, the survivors, had allowed the Germans to lead us "like sheep" to the slaughter.

When I joined the Israeli Army, for example, I was surrounded by many so-called heroes. They were all tough guys. They constantly boasted about all the Arabs they had killed. And they really looked down on me for having been a camp victim.

When I tried to tell them what I had gone through, my army buddies would ask, "How could you have let the Germans do this to you?"

It's taken me years to come up with the answer.

When we were liberated from Auschwitz, I remember the Russians captured some five thousand Nazi SS men. I saw these two "Mussulmans"—Jewish inmates, thin as skeletons, who had been left to die—take the guns away from these SS men and start shooting. And you know what? Not a single one of the Nazis even tried to run away. I watched as the Germans sat, awaiting their turn to be killed. There were only these two frail little Mussulmans. Yet the Germans sat there—like sheep.

As we, the Holocaust survivors, became stronger, we began speaking up. "What are you talking about when you say we went 'like sheep'? Wasn't there an uprising in Treblinka? Didn't we fight in the Warsaw Ghetto? And what about the Partisans?"

And today, when I am asked that question, I tell people it doesn't matter whether you're Hungarian, Polish, Jewish, or German: If you don't have a gun, you have nothing.

The Jews who were brought by cattle car to Auschwitz weren't told they were going to a death camp: they believed they were going to be working. They had not worked in years. They were hungry, and they wanted to eat. They thought that by working for the Germans, they would have food and money, and they'd be able to survive until the war ended.

That's why they went quietly—that's why they didn't cry, or shout.

And then Dr. Mengele would tell them, "Please take off your clothes because you need to take a shower." And off they went into the gas chambers, very quietly. Everything was done very quietly. When was there even time for an uprising?

There were people inside the camps who found ways to smuggle out letters to relatives and friends describing what the Germans were doing.

But absolutely no one believed them.

❧

ZVI THE SAILOR:

One day, the Israeli government selected me to join a group sailing to Germany to collect goods for Holocaust victims. The German government had decided to make some special reparations, and we were to pick up their gifts and bring them back to Israel.

We were very excited; everyone on the ship was a concentration-camp survivor.

When we sailed into the port of Bremen, we were greeted by German Coast Guard officers. We noticed immediately they were wearing the same uniform as the SS. It was exactly the same—the jacket, the trousers, even the cap.

They were very nice to us, these Germans—but we hated them anyway. Our government should have known better than to send a bunch of Holocaust victims back to Germany.

They would greet us every day by saying, "Good morning," but what we heard was, "Dirty Jew."

One night, we went to a bar in the port. We got roaring drunk and started a brawl. The police arrived, arrested us, and dragged us to the police station. We were enraged. We shouted, "Fine: We'll take over the police station." Then, we started fighting with the cops, too. We were ready to go to war with the German police, using the same tactics we had been taught in the Israeli Army.

The Germans were totally dismayed. They didn't know what to do. And to their credit, they acted like perfect gentlemen. They did their best to behave themselves. It was clear they didn't want any trouble with Jews.

None of us were thrown in jail, even though we had done a lot of damage, both to the bar and to the police station. We were booked, placed on trial, and we had to pay a large fine.

Nearly twenty years after the war, West Germany was at last ready to showcase its repentance for the Holocaust: It would hold its own war-crimes trials. Preparations for the Frankfurt trials were encouraged by Fritz Bauer, a German Jew who was powerful in Germany's judicial system. Bauer, prosecutor general for the state of Hesse (a position equivalent to state attorney general in the United States), was the man who had secretly tipped off the Israelis on where to find Eichmann. He had done so only after concluding his own country would never launch a serious pursuit. Courageous and energetic, Bauer was attached to his native Germany, but also cynical about the Germans' willingness to atone for their terrible past.

But now, with Bauer's help, German atonement was actually taking place. He gathered some of the best lawyers in the country to take part in the trial. Their official mission was to ferret out and prosecute any remaining Nazi war criminals.

Mengele, now living on a small farm in Brazil, followed the proceedings with contempt. In a diary entry dated May 2, 1962, he complained about the "beginning of a new witchhunt in Germany." He was referring to efforts by prosecutors to locate defendants for the trials, which would not take place for nearly two years.

His writings suggest he was appalled that his beloved *Deutschland* could turn against its own. He had never stopped loving and missing Germany, but he was enraged at the way it was treating men like himself—loyal soldiers of the Third Reich. Mengele still longed for what he called in a letter to his old school friend Hans Sedlmeier "the incredible zest for life of the German nation under Adolf Hitler."

In the same letter, he delivered a stinging indictment of postwar Germany. He noted that "ninety percent [of the German population] were made to suffer and feel guilty." Any people with Nazi ties had been classified by the government into different categories of culprits: There were the "followers" and the "profit-seekers," the "guilty" and the "extremely guilty," and, finally, the war criminals. Depending on one's classification, "one paid one's dues and became a rehabilitated citizen." Mengele resented a system that parceled out guilt and innocence with such apparent precision and ease. He also perceived its fundamental hypocrisy. "Later, those who had formerly occupied high positions got their retirement income again (slightly lowered) or jobs in private industry," he wrote Sedlmeier.

PETER SOMOGYI:

I left England for Canada, and there I was able to get a good job very quickly—with a German firm. It was called Bosch, and although it had been one of the largest industrial companies before the war, it had not used any slave labor.

The company's German executives knew I was Jewish when they hired me: I told them straight out. I also told them I was a concentration-camp victim. I even showed them my tattooed number. It turned out there were a couple of other Jews working there, too. Most of the German employees were very careful about what they said around us. Many of them had emigrated to Canada; they had probably been Nazis.

The firm was very good to me. I was promoted quickly. I was transferred from the Toronto office to Montreal, and made manager. I was able to send for my father, who was still in Israel. I even got him a job at Bosch, also.

In Montreal, the office manager's husband was also German. Out of the blue, one day, he formally apologized to my father for having been a Nazi during the war. He told him how much he regretted his past actions.

I continued to be promoted. My bosses liked my work, but I knew even then I would probably never reach the top, because I wasn't German.

One day, one of our regular German customers dropped by. We started chatting. He said he had a Jewish boss and that he didn't like him very much. He complained his boss didn't want to give him a raise.

He was confiding in me for he assumed that because I worked for Bosch, I was German. He had no idea I was both Jewish and a Holocaust survivor.

The more he talked, the more he made anti-Semitic remarks. He finally said something about "those Jews—isn't it a pity Hitler didn't finish the job."

That's when I lost my temper. I grabbed him by his shirt and told him, "You're talking to the wrong man. Get out, and don't you ever come in here again."

Then I went to my German boss and told him what I had done. I said, "Customer or not, I had to throw him out—you can fire me if you want." But my boss was behind me all the way. I continued to do well in the firm.

In Mengele's eyes, the "new" Germany forgave those it wanted to forgive, while making scapegoats out of a small number of others. He may have been thinking jealously of his former scientific colleagues who were now successfully reintegrated into German life, foremost among them his old mentor Verschuer, now professor emeritus at the University of Münster. Any hint of his past crimes had been conveniently erased.

How stark the contrast to what had been done to Mengele. Even after all these years, still "one revived the guilt of certain people," he wrote in the same letter to Sedlmeier, alluding to efforts by Germany to bring Nazis to book. He seemed to see a parallel between his own plight and that of Germany. Quoting former German president von Hindenburg, he observed, "It is easy to tread on a dead lion."

Life on the remote Brazilian farm was lonely and miserable. He was posing as a Swiss emigré named Peter Hochbichler. His companions were a Hungarian couple named Geza and Gitta Stammer. His old friend Colonel Rudel had asked another acquaintance, a fanatic Austrian Nazi named Wolfgang Gerhard, to find people who could protect Mengele. Gerhard had recommended the Stammers, who had been avid fascists prior to leaving their native Hungary during the Russian invasion of 1956. Even if they learned of the lonely SS doctor's identity, Gerhard gambled they would be sympathetic. A deal was worked out. The Stammers were to care for "Hochbichler," about whom they were told only that he was a Swiss exile. Mengele family money helped pay for a small farm the Stammers would "own." Mengele would assist in managing the farm—at no salary—a deal reminiscent of his days on the Fischer farm in Mangolding.

Mengele went to live with the Stammers in 1961. The Stammers were amiable enough, especially the wife, Gitta, but they were clearly on neither Mengele's social nor his intellectual level. After the dazzling Nazi company he had kept in Argentina and Paraguay, it was hard to get used to spending his time with a couple of uneducated farmers. With little else to occupy him, Mengele concentrated on his memoirs, a project that required painstaking effort on his part. That first year, he frequently felt blocked. "Beginning of memoirs difficult," he says in one diary entry. In another, he explains he has rewritten a portion of his autobiography, and "I am still displeased with it."

Mengele was unhappy in Brazil, a country whose multiracial population he could not like. Middle-aged, tired, and grumpy, he undoubtedly felt very much the outsider in this land of the young and exuberant. At one point, as Mengele was arduously transcribing his innermost thoughts, the noise and revelry from local villagers' *Carnaval* festivities disturbed his concentration. *Carnaval*, the traditional holiday held the day before Lent that Brazilians celebrate with special joy and ardor, had also been an occasion for parties in the Günzburg of his youth. Then, the young Beppo was seen at all the Mardi Gras balls. But now, he only expressed loathing at the joie de vivre of his Brazilian neighbors. Deeply bothered by the revelry, he complained in his diary that the noise hampered his work on his memoirs.

Although he was only in his early fifties, the once-promising young doctor was turning into a crotchety hypochondriac, writing daily reports of how he felt, and constantly worried about developing some dread disease. "Strong migraine and attack of aphasia," wrote Mengele on January 24, 1962. (Aphasia, the inability to speak or understand the spoken word, would have been an alarming ailment for the voluble Auschwitz doctor, and possibly a forewarning of the stroke he was to suffer years later.) The next day, he reported he was "very tired" and had an earache. On the twenty-sixth, he was "better" but his sleep had been disturbed by a very bad dream. He did not describe the dream, beyond saying he had been shouting.

MIRIAM MOZES:

After I got married, my husband noticed that I cried in my sleep. I had terrible dreams, and I cried so loudly, my husband had to wake me up to get me to stop.

❧

HEDVAH AND LEAH STERN:

The nights were especially bad. We would feel so frightened—as if we were being persecuted.

❧

MENASHE LORINCZI:

My wife told me I cried every night—I cried and shouted that people were trying to kill me.

❧

JUDITH YAGUDAH:

I have nightmares. My husband wakes me up and says, "Judith, Judith, you were shouting and crying in your sleep."

❧

VERA BLAU:

The children of the Holocaust did better than the adults. We did not realize the horror. We didn't even have to work. There was nothing for us to do at Auschwitz except to play.

After the war, I went through several stages. At first, I was very listless, very apathetic. I never cried.

Then, I cried. I cried frantically. I could not stop crying.

At the end of January 1962, there was a reprieve for a week. In one surprising entry during this period, Mengele seemed positively joyous, describing in detail the splendid Brazilian summer. He noted the blossoming flowers outside his window—red hibiscus, yellow lilies, and geraniums over an expanse of green fields. By early February,

Mengele felt well enough to spend several hours in the sun "reading poetry out loud."

But then the litany of complaints quickly resumed. Mengele was plagued, it would seem, by every possible discomfort, from headaches and insomnia to heart palpitations and dizzy spells. What probably contributed to his real and imagined ailments was an overwhelming sense of anxiety.

The references to his physical and emotional symptoms are among the few intimate details the author of the voluminous diaries included in them. The bulk of the documents Dr. Mengele left behind were impersonal and mundane, containing exhaustive references to the most trivial details of his daily life. There are notes on the weather, the servants, the coffee he drank that morning, the fence he painted that afternoon. The contents of dozens of notebooks that supplemented the diary entries are equally banal. Literally hundreds of pages are devoted to discussions on the genetic traits of birds and monkeys, the composition of blood, and an occasional venture into philosophy. These documents, which Rolf inherited after Mengele's death, were made available to a select number of German publications in 1985, and examined by scholars anxious for clues as to the Nazis' thoughts about Auschwitz. But even a close reading of the material turns up no obvious references to Mengele's time at the concentration camp. Rolf may well have discarded any material that cast his father in a bad light. But there is also the possibility that the Angel of Death couldn't write about that period of his life, either from fear of being discovered or from a deeper desire to block out his own horrible past.

PETER SOMOGYI:

Those years after the camp, I did not think very much about Auschwitz: I wanted to wipe it out from my memory. It was not with me every day—I was able to block it out. For example, I never, ever talked about it with my twin brother. And so I was able to put the war behind me—perhaps because I never admitted my own feelings to myself.

❧

JUDITH YAGUDAH:

As the years passed, I tried not to think about Auschwitz at all. I pushed it back and tried to live in the present. I wanted to forget.

But it was there, inside of me, like a heavy parcel I had to drag along all the time. I could not be happy.

My mother was always sad. She constantly wore a tragic expression on her face.

She wanted to talk about Ruthie. She would remember things long forgotten. She wanted to tell me about them, but I didn't want to hear.

❧

TWINS' FATHER:

I tried to forget the Holocaust. I kept on telling myself how lucky I was. I never talked about what happened with my family—not even with Magda. We never spoke about the loss of her little boy.

We wanted simply to forget. My feeling was that what happened at Auschwitz was over and done with.

Yet, there were times when I would get very depressed.

❧

OLGA GROSSMAN:

As I see it now, having children really brought back memories. I spent nine months being afraid. I had a nervous breakdown and was hospitalized after every childbirth.

One of the doctors reminded me of Mengele. He was tall and wore a white coat—an image of Mengele.

He ordered shock treatments. I felt I was reliving the horror.

Instead of addressing the central episode of his life, Mengele filled his notebooks with lengthy critiques of books, which he ordered compulsively from Germany. In letters home, he included long lists of what he wanted to read, from the latest medical textbooks to treatises on genetics and natural science. He devoured complex works of philosophy, history, sociology, and the sciences. Spengler, Konrad, Heidegger, Kierkegaard, even the Jewish intellectuals Martin Buber and Hannah Arendt (herself a German war refugee), are cited in the journals and notebooks. It is far from certain that Mengele understood these books—only that he knew they were what a well-educated German ought to read.

After perusing Camus's *The Fall*, for example, Mengele decided it was terribly overrated. Its author, the renowned existentialist and

Resistance fighter, was contemptible in Mengele's eyes. The Auschwitz doctor decried the "pessimists" of the existentialist movement who were both "Nazi-haters and communists." He delivered a blistering critique of *The Fall*, perhaps Camus's greatest novel. The book tells the story of a powerful, arrogant lawyer who watches a woman drown one night and is forever haunted by the fact that he didn't even try to save her. The lawyer's failure to act at that critical moment destroys his self-image. He renounces his profession and spends his days in a sleazy bar, rehashing his sham life.

But Josef Mengele, killer of thousands, could not conceive of feeling remorse for allowing one single woman to die. He did not even grasp the ironic parallels between the fate of Camus's hero and his own dramatic fall—the fact that he, too, was spending his life in a run-down hovel, rehashing the past.

Throughout the mundane descriptions, the banal observations, Mengele's journals betray an intense loneliness, a longing to return to Germany and what was left of his family. But his homesickness for *Deutschland* was always tempered by the bitterness he felt because his country had rejected him. "You have made it very difficult for some of your sons, holy Fatherland, but we will not desert you, and always, always love you," he observed.

Mengele's isolation was relieved only by occasional visits from trusted friends. But since not many people knew of his Brazilian hideaway, these were few and far between. Mengele, always cautious, seems to have broken with the Nazi circles he had frequented in Buenos Aires and Paraguay. Narrowing his circle of friends probably kept him safe, but it also made his predicament all the more desperate and galling. He was apparently determined not to permit Eichmann's fate to befall him.

Mengele's 1962 diary suggests mounting tension as the year wore on. The man who had whistled gay tunes amid the carnage of Auschwitz was unable to stand the isolation of his Brazilian hideaway. "Very bad headache" . . . "Very nervous" . . . "Condition worsened . . ." Even reading, his favorite pastime, was an ordeal when he suffered from one of his migraines.

Sometime later that year, the Stammers sold the property and moved to a farm closer to São Paulo. The new farm was in the town Serra Negra, and it, too, was purchased with the help of Mengele family money. From a perch atop the eight-foot tower he ordered,

Mengele would anxiously watch the roads and fields, on the lookout for any unwanted visitors. His dogs were always close by him, barking and growling, wherever he went.

MIRIAM MOZES:

When my daughter was little, and I took her for walks, I would get very frightened if we saw a dog. Since Auschwitz, I was terrified of dogs, and I passed this fear onto my child. She learned to think they were very dangerous.

One day, my husband took her out, and he noticed how she jumped when she saw a dog. She behaved exactly like me. He told me, "This is not good. We have to get a dog for her to get over her fears." I refused—and I continued to be afraid of dogs.

I would dream dogs were barking at me, and pursuing me in the night. They were terrible dreams. I would cry so loudly that my husband and children would have to wake me up.

Like a macabre parody of Proust, the gentle French novelist who spent the last years of his life in a cork-lined room recapturing his youth, Mengele's only source of pleasure now came from remembrances of his past. One day on the radio, he heard an old German hymn, perhaps from his days as a church-going Catholic in Günzburg: "Great God, We Praise You." He wrote in his diary, "[W]hat a nice memory." Another day, he read a book about Hungary and recalled his travels there as a young man. "A lot of old memories: pity that one did not get to know this country and the people better."

But whatever nostalgia Mengele may have felt for the Hungary of his youth, he began to treat the Hungarian couple he lived with poorly, subjecting them to ugly outbursts. The Stammers came to dread and fear their erratic "houseguest."

Gitta's husband was a sailor, frequently away at sea, and she was left alone for long periods with Mengele. She tried her best to create a pleasant atmosphere for him. An excellent cook, she would fix him his favorite dishes, chat with him, and even play the piano for his enjoyment. There were even rumors they were lovers. But instead of being grateful for her warmth and company, Mengele grew even more tyrannical, she later claimed in press interviews. He also became domineering and irascible toward the servants and other farm help.

On one level, Mengele's mistreatment of the Stammers reflected his disdain for Slavs, a people the Nazis had considered only slightly superior to the Jews. But his rudeness wasn't only a reflection of his sense of social superiority. He had treated his Aryan wife Irene much the same way, ordering her about and subjecting her to his jealous diatribes. The pent-up rage Mengele had probably always felt was manifesting itself in his relationship with the Stammers. Although he was both lonely and deeply dependent on them for protection, Mengele still wanted to feel in control. The fact that his family's money had paid for the farm gave Mengele economic power over them, power that he used indiscriminately. He even gave the Stammers advice—if not absolute commands—on how they should bring up their young children: He told them they should be far stricter with them, and discipline them more often.

Alienated from the Stammers, Mengele turned to his new friend, Wolfgang Gerhard, for friendship and support. Gerhard was Mengele's only friend in his Brazilian retreat. Gerhard, who had lived in Brazil since the end of World War II, could advise Mengele on the country and its inhabitants. He was also as fanatic and dedicated a Nazi as Mengele. This was their most important bond, since aside from a shared devotion to Hitlerian ideals, the two men had little in common. The crude, lower-class Austrian was certainly not the type of person Mengele would have associated with back home. But in the wilderness of his South American exile, Mengele was glad for any friends he could get. Besides, the admiration verging on adoration that Gerhard lavished on him was salve for his battered ego.

While still a teenager, Gerhard already a passionate devotee of Hitler, signed up for the German Navy. He named his oldest son Adolf, and long after the war had ended and he had left his beloved Austria, he continued to lament the loss of the glorious Third Reich. In his house were numerous souvenirs of the Hitler era, including medals and uniforms. Gerhard's wife, Ruth, was almost as devout a Nazi as he. She once gave their Austrian landlady a gift: two bars of soap still in their original 1943 wrappers. They were "Jew soaps," she'd said—made from the fat of Jews killed in the concentration camps.

Another of the Gerhards' prized possessions was the large swastika that topped their Christmas tree every year. "You always have to take good care of a swastika," the old Austrian liked to say. Truly, Mengele could not have met a more ideal protector. For Gerhard, watching over Mengele was a labor of love.

Although Gerhard had introduced Mengele as Peter Hochbichler, Swiss exile, sometime in 1963 Gitta Stammer realized his true identity. She stumbled on a newspaper containing a picture of someone with a striking resemblance to her houseguest. Mrs. Stammer recalled that when she confronted him with the photographs, "Hochbichler" turned white and quickly left the room. Later that day, he returned and admitted that he was the notorious Dr. Mengele of Auschwitz.

PETER SOMOGYI:

When I first met my wife, she immediately noticed the number tattooed on my arm. "Were you in Auschwitz?" she asked me. I didn't answer her. I didn't want to talk about it.

Her father was also a camp survivor, and he had a number on his arm as well. She knew what it meant. I brushed her off, and we both decided we didn't like each other very much, that first date.

But we went out again with another couple a few weeks later. It was to see the opera Don Giovanni. *We realized we both liked classical music. Even though she was ten years younger than I, she was very mature—perhaps because of what happened to her father.*

Nevertheless, I continued to refuse to talk about the war with her. Once, her father asked me about my experiences at Auschwitz, but I brushed him off as well.

It was a short courtship. We had met in August, we got engaged in December, we married the following June. I wanted to settle down. I had moved around too much from place to place.

After our honeymoon, when we returned to our new apartment, my wife said to me, "Now, will you tell me, please, about your past? I want to know what happened."

And so I told her in a nutshell about my experiences as a Mengele twin at Auschwitz. But only in a nutshell.

We had been married two weeks. We were starting our new life together. I said to my wife, "I have told you now what happened to me. Please don't ask me about that ever again. I never want to discuss it again."

Afterward, Mengele refused to talk about his past with the Stammers. According to Gitta, he didn't even like to mention the Second World War. Intensely paranoid after his secret was out, Mengele viewed any unknown visitors to the farm with suspicion. He was always asking her about the guests, who they were and what they wanted. Ironically, visitors to the Stammer farm who met "Señor Pedro" tended to be charmed by him. Mengele was exceedingly polite to strangers, and, when he felt he could trust a person, friendly and expansive. Gitta, who endured his constant harping in private, was mystified. It seemed to her that Josef Mengele had two sides, "one for strangers, and the other when he did not need to dissimulate."

HEDVAH AND LEAH STERN:

We are like actresses. Both of us hide our true feelings. On the outside, we are laughing and smiling, but inside, everything is rotten and dark, and will remain so until the end of our lives.

To new acquaintances, Mengele was the lovable Beppo, a sunny boy grown into a delightful if somewhat eccentric old man. Once the guests were gone, he behaved like a tyrant, given to cursing fits and temper tantrums. He was furious if he didn't get his way. Gitta perceived the fundamental split in Mengele's personality. She saw very clearly the duality, and it frightened and puzzled her. She was unable to fit the "twin" sides of Josef Mengele together. What she didn't know was that Mengele had always been like this. At Auschwitz, he would smile and act kindly, then fly into a rage at the slightest provocation. And here in Brazil, he had lost all of his power and none of his madness.

Yet, because Mengele made such a good impression on visitors, the Stammers had a hard time convincing their friends how irrational their dapper houseguest really was. Later, they would recall how they would agonize over the behavior of this strange old man, who seemed both to yearn for their friendship and to go out of his way to torment, insult, and abuse them.

The Stammers were coming to realize that they were stuck with their very difficult houseguest. They couldn't get rid of Mengele, nor could they persuade him to behave in a more cooperative manner. Whenever the Stammers complained and suggested a parting of the ways, Gerhard would arrive to deliver a stern lecture on what an "honor" it was to lodge the famed SS doctor. When that approach ceased to work, Gerhard resorted to threats. He suggested harm might befall them and their children if they dared to throw Mengele out of their home.

Eventually, even threats stopped working, and sometime in 1962 the Mengele family learned of the Stammers' unhappiness with their living situation. They quickly dispatched their best factotum, Hans Sedlmeier, who still enjoyed a top position in the firm, to smooth matters over. Clearly, the family wanted to continue the current arrangement. It was inexpensive, and they didn't have to worry about Josef's safety.

In Brazil, Sedlmeier listened sympathetically to the Stammers' complaints. Then, believing there was only one way to handle the distraught Hungarians, he offered them more money. Sure enough, the relationship was salvaged, at least for the time being. After Sedlmeier's visit, life settled down on the farm. Mengele continued to work on his memoirs. Progress was slow, perhaps because Mengele, sensing his future was bleak, preferred to linger over the past: He devoted scores of pages to describing his birth.

But even as Mengele quibbled with Gitta Stammer over the servants, or agonized about what word to use in his "memoirs," a storm was brewing in Germany. During the three years spent preparing for the Frankfurt trials that got under way in 1964, prosecutors had become convinced of the enormity of Mengele's crimes. The German government was prompted to intensify its search for him. Perhaps the most dramatic example of Germany's dedication to the hunt was their decision to have their ambassador to Paraguay intercede. In February 1964, Ambassador Eckart Briest requested an audience with President Stroessner. In a rare display of diplomatic passion, he demanded that Paraguay turn over the infamous death-camp doctor. Stroessner was so infuriated, he threw the ambassador out of the country, creating a minor diplomatic crisis. Because of his close ties to Nazis such as Rudel and Von Eckstein, the Paraguayan strongman was in a position to know—or find out— Mengele's whereabouts. And because foreign citizens were watched very closely, Stroessner must have had available precise knowledge of Mengele's trips in and out of Paraguay. But Stroessner was unwilling to betray a Nazi.

The Germans were determined that he be brought to book. But even without Mengele at the stand, this trial yielded ample testimony about his conduct at Auschwitz. The revelations by former inmates who had worked with Mengele elicited headlines—the first time since the end of the war that Mengele had received so much notice in his own country. Even in his own *Deutschland*, Mengele was now vilified, an object of loathing.

At the trial, an Israeli doctor named Mauritius Brenner told the German court how his twin children had been put to death by Mengele because they were not identical. His story was confirmed by the Auschwitz pharmacist, Viktor Capesius. Before a horrified court, Capesius recalled bringing the Brenner twins to Mengele, who was in an irritable mood. On that day, Mengele didn't want to be bothered with any fraternal genetic specimens. "I have no time now," said the Angel of Death, after throwing a glance at the bewildered children. They were promptly taken away to be gassed. Other witnesses described Mengele's selections, his affinity for experimenting on cripples, dwarfs, and, above all, twins.

Germany persisted in its efforts to find the fugitive, and offered a small reward for Mengele's capture. Fritz Bauer, the prosecutor who had helped Israel find Eichmann, made headlines when he alleged that Mengele had been spotted in Asunción in the company of Martin Bormann. Bauer insisted that Bormann, who was supposed to have died in Hitler's bunker in 1945, was in fact alive and well in South America. Bauer alleged that Hitler's most trusted assistant was good friends with the Auschwitz doctor.

But as German authorities searched far and wide for clues that might lead them to Mengele, they seem to have overlooked connections much closer to home: Hermann Langbein, who had forced Germany to reopen its Mengele investigation in the late 1950s, was involved in a new battle. Langbein hoped to persuade the University of Munich as well as the University of Frankfurt to revoke the degrees and doctor's license they had once bestowed on Mengele.

In the spirit of the Frankfurt trials, both institutions seemed receptive to Langbein's demands. The man who had killed hundreds of thousands of people certainly did not deserve the title of "doctor." The University of Frankfurt, embarrassed about its past as a Nazi academic haven, promptly agreed. The University of Munich said it would follow Frankfurt's cue.

But even as Frankfurt prepared to revoke Mengele's degree, Martha—still legally his wife, even though they were separated—launched a formal protest. The battle pitted the universities against a barrage of tough, highly paid Mengele family lawyers. In retrospect, it seems probable that the family was being egged on by Mengele himself, that Martha's prominent role was due to pressure from her estranged husband. Incredibly, however, German authorities never thought to investigate Martha or the rest of the family at this time, and to demand to know Mengele's whereabouts.

Indeed, although the Mengele family adopted the party line after the war that Josef was "missing" or dead, they came to realize there was no need to continue the pretense. Until the Frankfurt trials in the early 1960s, postwar Germany had not the slightest interest in finding, let alone trying, the Auschwitz doctor for war crimes. In a way, the mysterious Dr. Mengele was not so elusive after all, at least, not for many years. The townspeople of Günzburg knew that their Beppo was alive and well and living in South America. They knew of his divorce from Irene, and his bizarre marriage to Martha, and they delighted in gossiping about that union when it, too, failed. But no one was especially troubled by the knowledge that the infamous war criminal was still at large. None felt a need to come forward and tell all. More important, no government authorities—not the Germans, not the Israelis, not even the Americans—had ever bothered to question them closely until the mid-1980s, and by then it was too late.

Gently sidestepping the question of whether the aggrieved party was dead or alive, Martha and the family lawyers kept doggedly appealing the Case of the Diplomas. They argued Mengele's degrees couldn't be repealed because they had been granted before the war—and hence, before Mengele had committed his alleged crimes. But the University of Frankfurt refused to budge, and ultimately it prevailed.

The loss of his credentials must have been unbearable to the exiled Nazi. He could no longer harbor any illusions about returning to his homeland and his former life. Never again could he mention his prized doctorate, never could he try to assume a position in academia, his lifelong dream. Mengele's degrees were the last remaining testament of his former greatness, a reminder of his life before all went awry.

MOSHE OFFER:

After I left Blumenthal for the second time, I went back to school and resolved to change my life.

I decided to become a chemist. I worked very hard at my studies.

I also got married again, to an Iraqi woman. I was anxious to have a real home. We settled down and had children very quickly. We had five daughters, one after the other, and at last the son I had always wanted. I named him "Shai," which means "gift" in Hebrew.

I got my break when I was offered a job in a film laboratory in Tel Aviv. This laboratory made photo-developing film.

I was still very inexperienced, but this elderly German professor who worked there took me under his wing. He ran the laboratory, and he was a leading expert in his field. He taught me day and night. He gave me hundreds of pages of notes and formulas to study.

But this professor turned out to be corrupt. He was stealing silver used to make the film. The owners of the laboratory found out about it. They let him steal and steal and steal, until they built a case against him. One day, detectives came in, and they caught him with six kilos of silver.

They fired him and gave me his position. I was placed in charge of the laboratory. A lot of people worked under me. And people all over the country asked me for advice. I was an expert, after all—the old German had taught me everything he knew.

Because of my expertise, I became known throughout Israel. I got a lot of job offers. One day, Israeli television offered me a job. They offered me a lot of money. And so I left the laboratory to work for them.

❧

JUDITH YAGUDAH:

After years of struggle, my husband decided to start his own business. My uncle helped us buy an apartment for my mother—in the same building. She lived on the first floor, and we lived on the third floor. But at least we lived on our own. It was wonderful.

I helped my husband a lot in his new business, an electronics company. I encouraged him. I would even type letters for him, because he had nothing then, neither a secretary nor his own office. And little by little, he became successful. Our lives started to improve. We were even able to buy our own house.

We left my mother in the apartment in Netanya, and we moved to another town. I wanted to be free of her influence. I didn't want to live in the past.

With the Frankfurt trials, Germany atoned for its sins before the rest of the world. Once the proceedings were concluded, the Germans never launched as intensive a hunt for Mengele—or any other war criminal—again. As Mengele himself pointed out, many prominent Nazis continued to be rehabilitated, and resumed positions of importance. His old mentor, Verschuer, retired from the University of Münster in 1965 after a decade of honors and high praise. He had succeeded in building one of the largest genetics institutes in West Germany. When he spoke publicly on the subject of the Nazi era, he managed to sound as if he had only been an innocent bystander.

While Verschuer relished the glittering prizes heaped on him, his former protégé spent sleepless nights working on his memoirs. He also composed long letters to his family in Germany, as if to make sure they did not forget him. He was especially upset when his kinfolk failed even to acknowledge his birthday. When he turned sixty, he noted bitterly that Karl Heinz was the only one to send him a birthday card.

Peter Somogyi:

Every year on December 16, I would remember my dead sister's birthday. I would think to myself, "Now, she would be this old." I pictured what she would have been like if she had survived Auschwitz. I would imagine what would have happened if she were still alive, if we still had our family intact. But I didn't tell anyone my thoughts, not even my wife.

Mengele's greatest source of distress came from Alois, once his favorite younger brother. In the postwar years, Lolo had worked very hard to build up the factory. The natural heir to Karl Sr., Lolo successfully expanded the family empire. In the process, he also earned the reputation of being an exceptionally kind, fair, and honorable man. Both in and around Günzburg, Lolo was liked for his personal generosity and admired for his sound business sense.

At the start of his brother's exile, Alois had maintained cordial ties with him, even flying out to Argentina to visit. But later, Lolo deliberately sought to distance himself from Josef and barely kept in contact with him. A review of Mengele's letters to Alois suggest that he gave his older brother a hard time about his allowance, even though there was obviously money to spare. Since Alois controlled the family purse strings, Mengele was placed in the humiliating position of pleading for handouts from his younger brother.

Publicly, Alois Mengele continued to defend his sibling. But the townspeople of Günzburg say that over the years, Lolo became increasingly disturbed by the persistent stories about Josef's cruelty and sadism at Auschwitz. Alois evidently told Josef he had serious misgivings about what he had done during the war. According to the mayor of Günzburg, Lolo even chose to do his own research, going as far as to seek out witnesses who could corroborate his brother's version of events. But the mere fact that a family member would have doubts about him distressed Mengele terribly. The letters suggest there was a large rift between the two brothers who had once been inseparable, who had shared a passion for automobiles and pretty women, for swimming parties along the Danube and evenings in the Cafe Mader.

ZVI THE SAILOR:

A few years after the war, I stopped talking with my twin brother. We had always fought as children. We even fought in Auschwitz—and we continued to fight after the camp. There were a million things we didn't agree on. He didn't like my wife, for example. He never accepted her. Finally, he moved away to America, and I never heard from him again.

I sent him money, letters—but he never replied. I knew he got them, because I sent them through registered mail. I never got an answer.

One day, his wife, who comes to Israel frequently, dropped by to see me out of the blue. She said to me, "Leave your wife. Come to the States. We shall help you."

I told her, "Are you mad? You come to me after all these years and you ask me to leave my wife?"

In America, men do things like that—they leave many wives. What is that song, about buying a one-way ticket? The man who drops everything and leaves. Or the son who promises to keep in touch, and never sends his own parents a postcard? In Israel, family is much more important.

And even though I was always leaving Israel, I always, always came back.

❧

MENASHE LORINCZI:

My sister and I were very close after the war, but we drifted apart after she got married. Her husband was Hassidic—and she became very religious, too. I was no longer able to talk to her. She adopted all her husband's ideas.

We parted even more when she left for America. I wrote her letters—and she never even answered them.

❧

VERA BLAU:

My twin sister is my only surviving relative—and I love her, but with the years I found we had nothing in common.

When I get depressed, I have a "switch," and I can turn it off. I switch it off, and I stop feeling sad.

But my sister is always thinking about the Holocaust. She lives completely in the past. I can't bear to see her the way she is now.

There are two forms of theater—comedy and tragedy. I am a comedy, and my sister is a tragedy. She does not even like to laugh—while I enjoy Mickey Mouse.

And even though we both live in the same city, we hardly ever see each other.

When Alois developed cancer in the early 1970s, Mengele tried to mend fences with him. Upon learning of Lolo's fatal illness, Josef sent him a long letter expressing his deep sadness. He conveyed once again his dismay at the lack of regular contact between them. He complained he had only learned of his brother's illness "belatedly" and by chance. "Perhaps that is also part of my fate," he wrote mournfully, "but now, I only have the urge to communicate with you, dear brother."

Mengele confessed he had suffered greatly because of the "bizzare" conduct of his youngest brother. In a desperate attempt to effect a reconciliation, Mengele resorted to flattery, telling Alois what a wonderful job he had done running the factory. "As a reward for your exemplary lifestyle and great accomplishments, your town has bestowed on you the title of honorary citizen," he wrote in what was one of his last letters to Lolo. "It is a great honor, and I am very happy about the commemoration of your life's work, which has indeed been exemplary." Mengele noted how their own father had always thought Lolo could run the business better than anyone. He urged his brother to rest, and let his son and nephew take over the duties. "You should train your offspring to work in the business before it is too late," he warned.

Sensing the end was near, Josef thanked his brother for the help he had provided him over the years. He stressed how proud he was of Lolo and his excellence "as a speaker for our family, whose importance you have so greatly increased." The contrast between his brother's achievements and his own shambles of a life wasn't lost on Mengele. "I am especially happy over the distinction [you've earned] since I am part of the shadow," he sadly noted. The letter showed how much Alois's impending death had affected him.

MOSHE OFFER:

With the years, I found myself thinking more and more about my brother, Tibi. I pictured Mengele taking him away for the experiments. I remembered how sad he was at the end, when he could no longer walk. I would ask myself, "Why did I stay alive while he died?"

A twin brother is something very special. I have a very nice family, wonderful children. But I have no one to confide in. I feel that I want to say things I could only tell my twin. Instead, I keep a lot bottled inside of me.

With the years, I missed Tibi more and more. I kept wishing he were alive so I could show off my children to him. I liked to fantasize about the wedding he would have had; I wanted very much to see him happily married. I thought about the children he would have had.

Most of all, I wanted to introduce him to Shai, my son who is so much like him.

I would fantasize about this all the time.

Mengele's extreme sense of solitude was alleviated when he made some new friends. Gerhard introduced him to an Austrian couple, Wolfram and Liselotte Bossert, who were also die-hard Nazis. Wolfram had been a Hitler Youth leader during the war and had retained an abiding respect for the leader of the Third Reich. A locksmith by trade, he fancied himself a philosopher, and genuinely admired Dr. Mengele. Bossert considered it an honor to frequent the home of such an illustrious Nazi. His wife also enjoyed the company of the urbane, charming Auschwitz doctor.

The Bossert children "adored" Mengele, Liselotte would later tell the armies of reporters who swarmed around, asking for information about Mengele's years in hiding. Undoubtedly, the Auschwitz doctor was at his best with the Bosserts' son and daughter, as he had always been with young children. And they, in turn, Liselotte said, were thoroughly captivated by him. To them, he betrayed none of the mania that characterized his relationship with adults. Even as an old man, Mengele was more at ease with children than with grown-ups. The youngsters affectionately called him "*titio*"—little uncle—in a manner reminiscent of the twins and the Gypsies of Auschwitz. Both Liselotte and her husband considered the Angel of Death a good influence on their family. They had no qualms about their little ones spending time in his company.

His new friendship with the Bosserts provided Mengele with a badly needed social outlet. He was on intimate terms with the Austrian couple, and spent entire weekends with them in their beach house or exploring the countryside. At night, Mengele and Bossert, with Gerhard occasionally joining them, sat together chatting. They talked about politics, history, and modern-day Germany, which all agreed could not compare with the mighty Reich. Mengele's friends were dazzled by his wide range of knowledge, his ability to quote Greek and Latin texts, and, of course, the fact that he had once been a great doctor.

OLGA GROSSMAN:

I was in the hospital for months—but I hated the doctors in the white coats so much, I got worse—I rejected treatment. I wanted to die.

I missed my children, and they wouldn't let me see them. I wanted to go through the walls and run and find my kids.

The doctors were a little afraid of me. They didn't know how to approach me. I would strike at them, sometimes.

I was placed in another hospital, and there I met Dr. Stern. She was a young woman, herself a survivor of the camps.

She was very kind to me. I didn't feel I was just another patient to her. I felt she really cared. When I met her, I was twenty-five years old and I felt like a little girl. I believed she wanted to take care of me. She was like a mother to me.

My own mother was a sick woman. She couldn't cope with my problems. She had four children still at home, and she didn't have time for me. She couldn't come to visit me in the hospital as often as I wanted.

Dr. Stern was so sensitive to my feelings. For example, whenever she saw me, she removed her white coat, and stayed in her regular clothes. She never let me see her in white because she knew I was afraid of white coats. Dr. Mengele had always worn a white coat when he saw me.

Under Dr. Stern's care, I began to change. I started going out on walks. When I had been on many tranquilizers, I had lost a lot of weight. Now, I ate more, and even put on a few pounds. That was a seen as a sign of progress.

I had always been terrified of taking a shower. I would faint when I took a shower. But now, when nurses offered to help me, I refused: "I am going to take a shower by myself," I would tell them.

I finally made one request to Dr. Stern: to let me go home and see my children. "I promise to come back to the hospital," I told her. "I won't run away." Miraculously, she said yes. She told me she trusted me to go and return at a certain time.

I left for a day's visit. And I found that I wanted to come back. I asked Dr. Stern if I would ever be well enough to leave the hospital permanently. "Yes, yes—I promise you that one day, you will go home for good," she told me.

That's when I resolved to get well. I told myself, "I am going to get out of here. With Dr. Stern's help, I am going to get better." At the other hospital, I had wanted to commit suicide. Here, I wanted to come back simply to see Dr. Stern. I felt that she loved me—I felt I belonged to her.

Little by little, she got me to open up and talk about what had happened at Auschwitz.

10

THE SCHOLAR AND THE PREACHER

JUDITH YAGUDAH:

I don't like to live in the past. I don't like to be with other survivors. They are always dwelling on the past. "Do you remember this? Do you remember that?" Oh, no: I don't like that at all. I don't want to reopen old wounds.

I went back to my hometown of Cluj, once. It was my first visit since the war.

I spent a few days there with my husband. I walked around. I went to the house where I had grown up with Ruthie. I visited the last apartment we had lived in before we were deported to Auschwitz. It was very sad.

But looking for the past is always very sad, isn't it? Because we look and we find nothing—only memories and shadows. . . .

The 1972 Munich Olympics brought back fond memories for Mengele, who, as a young man of twenty-five, had attended the 1936 games in Berlin, a shining moment in Nazi history when Hitler had played host to the entire world. In a letter to Sedlmeier, Mengele urged him to send over special reports of certain events, in case they didn't receive adequate coverage in Brazil. "I am an old athlete," Mengele proudly reminded his boyhood friend.

But the games were ruined by the Palestine Liberation Organization's brutal attack on the Israeli team, an event that seemed to disturb even the old death-camp doctor. "Of course, this is not the proper method either," he lamented in his diary. Mengele apparently felt that there was a right way and a wrong way to kill the Jews. The random terror of the PLO's Black September faction possibly offended his scientist's sensibilities. As a eugenics doctor, Mengele had preferred a more systematic approach to eliminating the "inferior" race of Jews.

This murderous assault meant that Mengele had nothing to fear from the Israeli Mossad, which now devoted virtually all of its resources to infiltrating and rooting out Arab terrorist cells. Israel had neither the means nor the inclination to penetrate Nazi circles in South America. Besides, it was forging important diplomatic and military relationships with many of the Latin American states—including those harboring Nazis: Countries such as Bolivia, Uruguay, Brazil, Argentina, and Paraguay represented a potentially lucrative market for the Israeli weapons industry. These states were also proving to be solid allies at the United Nations, where a hostile clique of Arab sympathizers was threatening Israel's very existence.

But the greatest disincentive was the memory of Eichmann's capture. The Jewish communities of Latin America had been hurt and punished then. Any new operations would be sure to put them at risk. In Paraguay, for instance, while General Stroessner was willing to provide protection for a few Jewish families, he made it clear he didn't appreciate questions about Mengele. Israel was reminded time and again that Paraguay was its staunchest friend at the UN.

By the 1970s, the only insistent demands for Mengele's capture were voiced by Nazi-hunters like Simon Wiesenthal and others, who had made a career of hunting down war criminals, and who had stalked the Auschwitz doctor for decades with determination. From his musty office in Vienna, crammed with files and guarded by a sullen Austrian police officer, Wiesenthal liked to think he was keeping up with the elusive Angel of Death. He would hold frequent news conferences to announce he was getting closer to finding Mengele. The old Nazi-hunter constantly claimed to have caught glimpses of him on the run. A newsletter he published was replete with "updates" on Mengele's supposed whereabouts. It was only a matter of time before the infamous Nazi would be safely in the hands of authorities, Wiesenthal would confidently proclaim.

But while these reports helped keep the memory of the Auschwitz doctor alive, Wiesenthal could only have been a minor annoyance for Mengele and his family. He was usually so mistaken in his supposed sightings that he inadvertently helped out Mengele's efforts to remain in hiding. More than anyone, Wiesenthal helped foster the image of Mengele as a glamorous international fugitive, on the run between Argentina, Brazil, and Paraguay, with occasional jaunts to Europe, and harming anyone who came too close to him. Wiesenthal promoted and embellished the tale of "Nora Eldoc," an Israeli female spy who supposedly befriended Mengele in a South American resort, and ended up being thrown off a cliff, the victim of a mysterious "accident." Probably the single most repeated story involving Mengele, it was enough to strike fear in the hearts of legions of would-be Mengele hunters.

There was in fact very little glamour in Mengele's life, and a great deal of misery. The Olympics aside, 1972 proved to be rather a bad year. As a result of his obsessive worrying, which led him to chew nervously on his mustache, Mengele developed bezoars, hairballs in his stomach. This unusual condition, common to cats—and psychotics—was acutely painful and required surgery. Terrified of being discovered, Mengele approached a surgeon at a local São Paulo hospital using his alias, Peter Hochbichler. When the doctor ordered extensive X rays, Mengele insisted on having access to all of them. To the doctor's surprise, Mengele warned him not to make any extra copies of the films. The surgery was succesful, and Mengele made certain to retrieve all copies of his medical records.

According to Gitta Stammer, even after his convalescence, Mengele seldom left the house. The bulk of his time was spent reading and working on his memoirs, which by then had evolved into an autobiographical novel. Occasionally, he socialized with the Bosserts. When his brother Lolo finally succumbed to cancer in 1974, Mengele felt more alone than ever. His relatives in Günzburg were deluged with his lengthy letters.

With old age, Mengele's paranoid tendencies were becoming even more pronounced. His frequent letters home were written in a tortuous, virtually illegible script. He employed an elaborate code to disguise the names of people and places—but it was so arcane that he was the only one who understood it. His family, annoyed by the endless cryptic references, asked him to be more direct—and more succinct. They pointedly suggested that postcards would do as well in maintaining the relationship. But Mengele of course refused: He would have been robbed of his chief source of pleasure. His family then pleaded with him to type out the letters. Sedlmeier himself intervened to ask Mengele both to keep the letters brief and to space them out more. And Bossert sent him a note urging him not to use "so much code language in your letters . . . Make them so they are understandable to idiots, so we can all avoid the constant misunderstandings."

A special, top-secret arrangement had been worked out to get letters to and from Mengele. As with all sensitive matters involving Josef, Sedlmeier was the trusted intermediary, overseeing the scheme that allowed the war criminal to stay in touch with his family. Mengele would send his letters to a post-office box in Switzerland, and Sedlmeier made regular trips to pick them up, making sure they reached the parties to whom they were addressed. Any mail to Mengele also went through Sedlmeier, who sent it to a post office box in Brazil. This complicated arrangement kept up the connection Mengele so desperately wanted with his family. Alas, his nieces and nephews didn't feel any strong obligation to respond to the voluminous correspondence. Family members adopted Lolo's technique of simply not answering. The infrequent mail from home was a source of anguish for Mengele. There were times when his Günzburg relatives didn't even bother to retrieve the letters he sent them. "I always feel hurt when I become aware that no one is picking up my mail," he complained to someone code-named "Kitt." "Your neglect is a source of great pain to me."

Despite the family's cold shoulder, Mengele retained a passionate concern with trivia and idle Günzburg gossip. Like a spinster aunt with no life of her own, Mengele kept close track of the romances and little domestic squabbles, the weddings, pregnancies, and births in his boyhood town. "I heard a rumor that Dieter [Lolo's son] is going to be married," he writes to Sedlmeier. "Who is the lucky girl?" In the same letter, he sends warm wishes to a boyhood chum whose son has just become a doctor. "If I had to choose again, I would choose the same profession," the Auschwitz doctor blithely observes. "Congratulations."

MOSHE OFFER:

Happy occasions are the worst times of the year for me. Instead of feeling cheerful. I feel lost, remembering all the people who aren't here with me.

I become so depressed, I find myself thinking I would have been better off if I had died along with my brother at Auschwitz.

When I held my son's bar mitzvah, it was like a day of mourning for me. It was like a funeral.

I kept thinking about my dead brother, about all my family. "Why aren't they here with me?" I kept thinking. "Why am I all alone?"

❧

PETER SOMOGYI:

When my son was bar mitzvahed, it was supposed to be a very happy event for me—after all, he was my first-born son. But I could not stop crying.

We were all crying.

❧

JUDITH YAGUDAH:

Holidays and special occasions are the hardest. Here in Israel, everyone has a big family they can celebrate with. But we have no extended family.

My husband has no surviving sisters or brothers. And I don't have Ruthie.

My children would often ask, "Why don't we have cousins?"

❧

HEDVAH AND LEAH STERN:

During joyful occasions, we feel sad. If there's a bar mitzvah, or a wedding for our sons or daughters, we think of our mother instead. We remember her saying good-bye to us, and we get depressed.

But by 1975, Mengele had enough problems on his own home turf to worry too much over events in distant Günzburg: His loyal friend and protector Gerhard decided to leave Brazil, and Gitta Stammer resolved to kick Mengele off the farm. In Gerhard's case, both his wife, Ruth, and their son were dying of cancer. He hoped that by moving them back to Austria, there might be a better chance at a cure than in Brazil. Before he left, Gerhard gave his friend an invaluable gift: his identity papers. Sixty-four-year-old Josef Mengele, alias Peter Hochbichler, became fifty-year-old Wolfgang Gerhard.

But Mengele's other guardians, the Stammers, proved less charitable at the end. "You are not in any position to push me around," Gitta once told Mengele, alluding to his status as a wanted man. Although she didn't follow through on her veiled threats to expose him, she made it clear he would need a new home. She was going to sell the farm, buy herself a house, and devise a separate living arrangement for the death-camp doctor. Her days as a caretaker were over.

That same year, Mengele left the Stammers' residence and moved into his own house. Home for the Günzburg heir was a small, shacklike structure in one of São Paulo's more decaying neighborhoods. As a final favor, the Stammers had chosen the residence themselves, had purchased it, and rented it to Mengele. And with the profit they made from selling their own farm—originally purchased with money provided by the Mengele family—Gitta and Geza bought a splendid villa in an affluent section of São Paulo. It was a small enough reward for the years they had harbored the erratic Dr. Mengele.

Forced to live alone, Mengele felt more and more afraid. A small stroke that he suffered in 1976 added to his sense of helplessness. He was paralyzed for a brief period, and although he regained most of his faculties, his writings suggest he was deeply shaken. In his own apartment, cut off from the few friends he could claim, Mengele was a desperately unhappy man.

It was around this period that Mengele sought to improve his stormy, on-again, off-again relationship with his son. Rolf Mengele had become demonstrably cooler to his father than the rest of the family ever since he'd discovered his identity. One Christmas, when Rolf was still a young boy, Mengele had remarked on the difference between his son and his stepson, Karl Heinz. "The letter from Karl Heinz was very sweet—the one from Rolf was too factual." Now in his thirties, Rolf was still ambivalent toward his father.

In the 1970s, Mengele's letters to Rolf were replete with advice, stern warnings, and dire prophecies of failure. He displayed the same need to dominate that had characterized his relationships with friends—and with Rolf's mother, Irene. And not surprisingly, the more Mengele sought closer contact, the more he seemed to alienate his son. Indeed, it would seem that Mengele had become almost a parody of the engulfing maternal figure who had haunted his life. Sharp alternances in sentiment and mood characterized Mengele's correspondence with his son. In the space of a single letter, he could veer from affection to abuse, from a show of warmth to an outburst of rage.

MOSHE OFFER:

There are times I get so angry, I beat up my son. I get in such a rage, I don't know what I'm doing. I can't control my temper. I start throwing things and smashing furniture. I beat anyone who is near me.

Anything can trigger my rages—even a little thing, like a messy room in the house.

My second wife, Miriam, knew nothing about my past when she married me. She might not have married me if she had.

I used to beat her a lot. She was very frightened when I hit her or the children. But she knew I was not to blame.

When I calm down, I feel very bad. I tell my wife and my children I am sorry. I ask them for forgiveness. I beg them to understand.

As a long-distance father, Mengele managed to be both intrusive and critical. He probed for information on his son's personal life, even suggesting the types of women Rolf should date and marry. In one letter, he urged Rolf to let his intellect guide his relationships with women. "Rational observance" and not "emotion" should be used to select a mate. The old racial hygienist even threw in a brief discourse on genetics, listing the traits Rolf should look for in a wife. He cited, not surprisingly, good genetic stock as the number-one requirement. He urged Rolf to investigate a girl's family for its "rank and reputation." After that, her wealth naturally had to be taken into account.

But Mengele didn't limit himself to prying into his son's romances. He also injected his views on Rolf's work and career prospects, often expressing disapproval. His letters suggest he thought his son was shallow and superficial. "You and your generation have not been educated to survive," he wrote at one point. Mengele faulted Rolf for being too materialistic—for desiring money, cars, and other comforts.

But not all of Mengele's letters to Rolf were angry. In between the admonitions and advice, the incessant demands and harsh rebukes, he displayed a deep love for his son. In one letter, he reminisced about a day he was home taking care of Rolf, then just an infant. "There was an air raid, and I took you in my arms and ran to the bomb shelter," Mengele recalled. "The worrying and responsibility I felt for you as a child are still with me, as if it were yesterday," he wrote.

HEDVAH AND LEAH STERN:

When we became mothers, we were so overprotective, we never let our children out of our sight. Our lives were completely devoted to them.

For example, when they came home from school, they would sit down at the table and have their meals served on the spot—on the spot.

Occasionally, friends invited us to go out to dinner or to see a movie. But we would never go out with our husbands and leave the children with a baby-sitter. Either we went out or our husbands did—but we never left the house together.

❧

HEDVAH STERN:

One day, my son told me, "Look, Mom, you can go out with Daddy and leave me by myself—I'm a big boy now. I'll watch over the other children." He was ten years old at the time. But I simply couldn't deal with the thought of going out with my husband and leaving the children alone. I couldn't cope with that. I left the house—but I felt frightened the whole evening.

❧

OLGA GROSSMAN:

When my children were young, I would not let them leave my side. I needed to have them near me constantly. I was obsessed with them. Having them close by was like being able to breathe. After I saw them off to school in the morning, I collapsed.

There were days I wouldn't even let them go to school. I made them stay home with me, instead. They'd get embarrassed. "Mommy, we have run out of excuses," they would tell me. "What are we going to tell our teachers?"

But I didn't care.

Perhaps I was feeling guilty for all the time I had left them alone when they were small, and I had to be hospitalized. My conscience was killing me.

❧

JUDITH YAGUDAH:

I have two children—a boy and a girl. I worry about them constantly. I always feel afraid for them. I am a burden on them, I am sure.

❧

PETER SOMOGYI:

I worried constantly about my children. When they went away, I insisted they call home every day. I needed to have copies of their class schedules, their school schedules—I always had to know exactly where they were.

If they didn't call for a couple of days, I would tell my wife, "What's the matter with the children—let's call them."

I was always terrified I would be separated from them.

Mengele took out his frustration about his own failed career on Rolf, who was struggling to lead the appearance of a normal life. When Rolf informed his father he was not going to finish his Ph.D. in law, Mengele became livid. It wasn't enough that his son was already a professional, a trained lawyer assured of a good future. The man who had earned two advanced degrees, then endured the humiliation of seeing them taken away, seemed determined that his son should gain back the professional standing he himself had lost. A major argument erupted between father and son. In vain, Rolf tried to list the reasons he had abandoned his dissertation. He lacked the money to finish. He was no longer interested in the subject matter. He was tired.

Mengele promptly sent a scathing letter to his son. "I accept that you lack interest and are not willing to work—but I do not accept that you don't have enough money," Mengele wrote, apparently forgetting that Rolf had no claim to the family fortune. After the war, Mengele had signed papers relinquishing any share in the estate. Mengele went on to point out how foolish it was for Rolf to drop out at such a late stage, "especially as you have studied so many semesters, and have three years of training." Deeply upset, he remarked, "I do not have the proper words to express my feelings. I was very hurt. The doctorate was the only desire I ever had for you."

In his longing for his son to recoup the respect he had lost,

Mengele failed to realize that Rolf had inherited neither his intellect nor his fierce drive and ambition. In spite of the guilt his father made him feel, Rolf abandoned his doctoral thesis, and decided to practice law instead. He eventually opened a small law practice in Freiburg. It was a modest path for a man of modest ambitions. Dr. Mengele's son was to exhibit several of his father's traits as he grew older, from his charm to his obsessive-compulsive qualities, but the burning ambition, the deep need to achieve, was never among them. Rolf settled instead for a quiet existence near his mother, shunning both Günzburg and the Mengele clan. Irene herself had little or no dealings with the Mengeles. Whether she actively encouraged a relationship between Rolf and his father, or chose simply not to stand in the way, is unclear.

The relationship between father and son took a decided turn for the better in 1976, when Rolf became engaged to a blonde of good Aryan stock. It would be his second marriage; his first, to a pretty brunette who had been his childhood sweetheart, had lasted less than a year. Dr. Mengele was delighted with his son's new choice of mate. Almuth Jenkel was a stunning beauty, with long flaxen hair and big blue eyes—a modern version of the Hitlerian ideal. The large breasts and athletic look of the 1930s poster girls were out. Fashion now dictated a delicate, more slender type of Aryan womanhood, which Almuth embodied.

Best of all, she was a twin. "Mengele was both fascinated and delighted with the fact that his own son had married a twin," Rolf would later recall. Mengele grilled him about her genetic background. Evidently satisfied, he extolled the benefits of the union from another genetic point of view. "For the first time, one of our own has gotten his wife from north of the Main line," he exulted in a letter to Rolf. While his family came from southern Germany, Mengele, like other racial hygienists, believed the best genetic traits were to be found in the northern regions. "This movement north of the border is to be welcomed, and one can only expect the best from this new match."

Indeed, he told his son he hoped his daughter-in-law would bear twin grandchildren. "Even the characteristic of Almuth as a non-identical twin was of special interest to me," Mengele noted in his journals. And it is possible that Rolf was also trying to please this father he hardly knew but who had haunted his childhood and youth. If nothing else, it was considerably easier to wed the lovely Almuth than to finish a tedious dissertation.

OLGA GROSSMAN:

My daughter fell in love with a Canadian boy and decided to marry him and move to Canada. She settled in Calgary, near the North Pole.

I knew even then that she was running away from me—from my past. She had been forced to live with it since she was a baby. It was very difficult for her, and so I guess she did what she could to survive: She moved as far away from me as possible.

But once she got there, she realized the bond we had as a family was much too precious. She wanted to have children—but she couldn't bear the thought of her parents not knowing their own grandchildren. And so, her marriage suffered.

At about the time of Rolf's engagement, Mengele expressed a longing to see his son—a request he had made repeatedly over the years. The two had not met since the 1956 holiday in Switzerland, when Rolf hadn't known the handsome Uncle Fritz was really his father. Mengele's son hesitated; the relationship had been so troubled over recent years, as far as he was concerned, it might best be left to letters and photographs. Ultimately, pity and curiosity prompted Rolf to change his mind—pity for the father who was old and sick, and curiosity to better know the man who had been widely depicted as a monster.

Preparations for the trip to Brazil were hampered by the ambivalence both father and son felt about the trip. Josef kept dictating elaborate security precautions. Rolf kept postponing the date of the voyage. Once again, as with all family crises involving Mengele, Hans Sedlmeier stepped in. The perfect diplomat, Sedlmeier encouraged Rolf to follow through on his resolve by reminding him that his father was still badly shaken by a recent stroke: His last desire was to see his son. Sedlmeier told Mengele, meanwhile, that his son was of a different generation, and that he should not expect too much of him.

PETER SOMOGYI:

One day, I was leafing through Life *magazine, and I nearly jumped: There was Twins' Father—both his picture and an article about him.*

There was also an old photograph of my brother and [me], taken when we got back to Hungary after the war. We had mailed it to Twins' Father to let him know we had arrived safely.

My wife and I were living in America at this time. We had left Canada for the U.S. during the turbulent period of the Quebec Separatist movement. Many Canadian Jews had panicked and fled Montreal—including my wife's family. They felt very strongly that we should move, too.

As I read the article, I got more and more nervous. I was gasping for air.

The next morning, I called up the magazine. I asked them for Twin's Father's address: They didn't know it. I called up the Israeli embassy in Washington. "Can you give me his address?" I begged them. They informed me there were five Zvi Spiegels in Tel Aviv. "Give me all five," I said.

I wrote to all five of them. And sure enough, he sent me back a letter. He told me his daughter was living in Brookline, Massachusetts, and that he was planning to visit her. That's how we arranged to meet.

❧

OLGA GROSSMAN:

When my daughter separated from her husband, I decided I had to see her through the divorce. I decided to go to Canada and live with her for a while, and help her overcome her depression.

When I left for Canada, I was terrified. It was extremely difficult to leave my husband, my son, and Dr. Stern. But I said to myself, "My child needs me: Now is the time I can offer her my help."

I left my family for three months and went to stay with my daughter.

At last, a date for the trip was set, airline tickets were purchased. So as not to endanger his father, Rolf would travel with a false passport. In May 1977, Rolf flew from Germany to Rio de Janeiro and traveled from there to São Paulo. Although he was employing an alias—the passport he used was actually a friend's—he was constantly afraid he was being followed, and would unknowingly lead authorities to his father.

OLGA GROSSMAN:

I arrived in Canada at the worst time of the year. There was so much snow. It was beautiful—but it reminded me of Auschwitz. I settled in my daughter's apartment. When she left for work each morning, I would go do the shopping.

One day, I was caught in a terrible snowstorm. I was very frightened. That's when I told myself, "I am going to bring my daughter back home—where she belongs."

And that's exactly what I did. I brought her back to Israel.

❧

PETER SOMOGYI:

The day Twins' Father arrived, he called me on the telephone to arrange a date when we could meet. I was working, and it was hard for me to get away. We thought I should come up to Boston on the weekend—but I was extremely nervous. "I'll leave work now, and I'll come up and meet you," I told him.

I told my boss I had to leave, ran out of the office, got in my car, and started driving to Massachusetts.

When I arrived at his daughter's house in Brookline, I found a lot of reporters waiting for me outside. The neighbors had thought this would be an interesting reunion, and they had called the press. But I walked straight past the cameras into the house.

We didn't say a word to each other—not a single word. We simply hugged.

He had changed: It had been so many years. His wife was crying in one corner—his daughter was crying in another corner.

According to Rolf, the reunion was tender and sentimental. Mengele's "child" was thirty-three; Josef was sixty-six, ailing, and seemed much older. During Rolf's stay, the old Nazi was often on the verge of tears, so glad was he to be reunited with his son again. In interviews given years later, Mengele's son depicted his father as warm and loving. "You surely do not believe what has been said about me?" Rolf said his father asked when questioned about Auschwitz. It was then that Dr. Mengele swore "by my mother's eye" he had never "personally" killed anyone.

The rest of the trip was warm and familial. Their two weeks together were crammed with sightseeing trips, family gossip, and even some political discussions. At one point, Mengele and his son argued about the death penalty. Josef was in favor of it; his son was firmly opposed to it.

Dr. Mengele seemed anxious to be at his best. He proudly took his son around to meet his friends the Bosserts. They also ran into the Stammers during one of their outings. Although they were estranged from Mengele, the Hungarian couple was all smiles when they posed with Rolf for pictures. Mengele even played tour guide and showed Rolf the country he had reviled for nearly twenty years. In an effort to woo Rolf, he offered him the use of the lone bed in the house, while he slept on the floor. And perhaps it was to woo the public that Mengele's son recounted this incident over and over again, as if to show there was another, more human side to his father.

On the last day, the two took a trip to Bertioga Beach, where the Bosserts had their cottage. The weather was idyllic, and father and son relished their final hours together. Both sensed it would be the last time they would see each other. The next morning, Mengele insisted on escorting Rolf to the airport in São Paulo, in spite of the risks. But so as not to attract attention, the farewells were brief and somewhat impersonal.

LEAH STERN:

I was invited one year to attend a wedding in America. A dear friend of mine was getting married, and I decided I had to go.

My daughter accompanied me to the airport to see me off. She kept walking with me, but then we came to the gate, and the security officers wouldn't let her go any farther.

My daughter started waving and shouting, "Mommy, mommy." And I fainted. At that moment, everything came back to me.

I heard my mother crying, "Wait for me, wait for me," as she was separated from us at Auschwitz.

Rolf returned to Germany, having obtained his father's blessing to marry Almuth. Mengele had even used most of the pittance left of his savings to purchase a ring for his son's future bride. After Rolf's departure, Mengele deteriorated considerably. Often in pain, he spent much of his time in bed. He wrote obsessively. The daily entries in his diary became longer and longer, crammed with increasingly mundane details. Restless and unable to sleep, he read late into the night.

At a time when Mengele's life was filled with little more than complaints, a major change occurred: He fell in love. The object of his affections was a lowly Brazilian servant, nearly forty years his junior. Elsa Gulpian, who became his housekeeper in 1976, at around the time of his stroke, was worlds apart from his first wife, Irene, and she possessed neither the elegance nor the glamour of Martha. Young and impoverished, Elsa was the type of woman the Josef of old wouldn't have deigned to notice. Yet, she was sweet and charming and utterly servile. She was also one of the few people able to get along with him.

Elsa has recalled that the Auschwitz doctor—who she knew only as Señor Pedro—courted her with chaste ardor. As their relationship deepened, he pleaded with her to live with him. She refused; she wanted a wedding ring. But Mengele didn't consider telling his maid that he was technically already married. Although he had been separated from Martha for nearly twenty years, the marriage had never been legally dissolved. But there may have been other factors at play. However lonely and miserable he might have been, perhaps Dr. Mengele could not stoop to marrying a servant girl.

Their courtship didn't progress to a physical stage. It remained as pure as a soft white shawl Mengele gave her. Mengele admired Elsa's modesty and old-fashioned values, if only because he shared them. In his old age, the mass murderer of Auschwitz was adhering to a strict moral code. In his notebooks, he often criticized modern sexual mores, elaborating on what he felt was the "proper" conduct between men and women. Like a preacher, he railed against promiscuity and "perversion." The sexual liberation of the 1960s and 1970s made him intensely uncomfortable. "Sexual barriers prevent chaos and stabilize hereditary laws," he proclaimed in one entry. But he also observed, somewhat ambiguously, that it was fine for men and women to enjoy "a normal expression of passion."

Elsa, who had been raised a devout Catholic, continued to hope for a marriage proposal. But when she realized it would never be, the sensible girl became engaged to someone else. Mengele was devastated. Elsa's action was seen as a personal betrayal—even though he had prompted the breakup by refusing to marry her.

Mengele's increasingly strange compulsions distanced him even from loyal admirers such as the Bosserts. As Wolfram later confessed in a letter to the Günzburg clan, both he and his wife came to limit their contacts with the old Nazi, even though they liked him. Mengele, who ironically was on his best behavior with the Bosserts, was threatening to destroy their lives with his obsessive habits and arbitrary commands. He "demanded comformity of everyone," Bossert wrote. "Whoever didn't manage to keep his own head would be overrun by Mengele, until he lost his own sense of identity." That was why, Bossert said apologetically, a "constant tie" was impossible. "It was important for us to be somewhat distant . . ."

With Gerhard gone and the Bosserts restricting their get-togethers, Mengele became more homesick than ever. Even after twenty years, Brazil was a foreign land, which he viewed through the prism of his own boundless misery and despair. Mengele never learned to love the country, or its spirited inhabitants; he never appreciated its music or its culture. He spoke Portuguese, but only grudgingly. The sole concession he made was to admire the rich, dazzling Brazilian countryside.

Expressions of joy are rare in Mengele's diaries, especially the later ones, and nonexistent in his letters home. It is as if the aged Nazi wanted to elicit sympathy from his relatives by dwelling on his pain. Birthdays continued to go unnoticed: One year, he noted cryptically that "a visitor is making this sad day much more bearable." Another year, Bossert surprised him with a cake and hearty dinner —but "unfortunately, no letters." As old as he was, Mengele had remained a little boy inside, longing for the fanfare of his privileged youth, when March 16 had invariably meant lots of presents, love, and attention.

He tried hard to repair his relations with the new generation of Mengeles. Although he was close to Karl Heinz, he had virtually no dealings with Lolo's son, Dieter. He worried that they might harbor suspicions about his past. Because he couldn't go to Germany, he relied on trusted emissaries to keep the young family members firmly on his side. Once, Mengele asked Gerhard to visit his family in Günzburg and tell his relatives what a fine person he was. "I would like for them to meet you, and for you to tell them what is both good and bad about me," he wrote in a letter sent before the planned encounter. "This way, we can avoid the misunderstandings and the ignorance, the mistakes and the negative attitudes." Josef Mengele knew he was loathed and reviled by the entire world. Yet in his old age, he longed to redeem himself in the eyes of a few admirers in Brazil and his family in Günzburg. Dr. Mengele wanted to be looked upon compassionately. He portrayed himself as a lonesome exile, condemned to a life of penury and solitude. He was not an evil, ruthless murderer, but a victim.

Mengele's last complete diary, for 1978, reads twice as long as any of the others. A whiny tone pervades the manuscript, written in barely legible script. Every entry compulsively begins with a report on the day's weather, followed by a description of his latest physical ailment. In between the headaches and dizzy spells, stomachaches, earaches, and sense of fatigue, there are constant remarks about feeling depressed and defeated.

Mengele was racked with pain and tormented by loneliness, and that is perhaps the central theme of the late diary entries. But there is a subtheme as well, for woven into the text of the 1978 journal is his disapproving dissection of postwar global popular culture. He condemns the local obsession with South American soccer scores. He rails against a jazz festival he sees on late-night television as "musical schizophrenia." With contempt, he tells of a pair of visiting lesbians from the Galapagos Islands. With finality, he labels the brilliant German autobiographical war novel *Das Boot* as "pornography masquerading as politics." He watches the heavyweight title fight between Muhammad Ali and Leon Spinks, two black men, and dismisses it with disdain as "a symbol of the stupefied mass culture." Only one popular icon wins his admiration, an American actress in an obscure movie role, and while Mengele does not mention her film by name, he does write that it is the only picture he has ever seen starring Marilyn Monroe.

Mengele's insomnia now lasted for days and weeks on end, making him constantly edgy and exacerbating his physical discomfort. "I am losing hope that I will improve or heal in the future," he writes in one entry. "It is the fifth week of suffering through the days and nights without sleep."

MOSHE OFFER:

I sleep very little—two-and-a-half, three hours is a lot for me. I often find myself pacing the house in the middle of the night, while everyone else is sleeping.

I get up and go to another room so I won't disturb my wife. I get up at four o'clock in the morning, to be at work by five o'clock. I work very hard—twelve, thirteen, fourteen hours a day. I like to work hard—because then I can drive myself to a point of such exhaustion that I simply collapse and sleep a little bit.

I have terrible problems with my nerves. If I have the slightest problem at work, I can't sleep at all. I can only sleep when I am absolutely exhausted.

❧

HEDVAH AND LEAH STERN:

We have trouble sleeping. We can't seem to fall asleep, and we depend on pills. Usually, a Valium or two helps.

❧

MIRIAM MOZES:

I take medicine to go to sleep. Or else, I would get up and cry.

Mengele's unrelenting loneliness was relieved only by visits from Elsa, with whom Mengele had managed to preserve a friendship. Although happily married, the young Brazilian woman seemed to genuinely care for the old death-camp doctor. She watched over him, attended to his needs as best she could, and provided him with news about mutual acquaintances. Mengele thoroughly enjoyed talking with her. "Her childish, primitive opinions are touching," he observed in his diary.

The rare times Mengele did venture outside, his neighbors saw the same elegant old gentleman they had known for years. Just as he had in his youth, he continued to pay meticulous attention to his appearance. Through the shabby streets of São Paulo, Mengele walked around in a Burberry raincoat. His matching Burberry hat was pulled low over his forehead, as a security precaution.

Although he was ailing, Mengele worked urgently to finish the autobiographical record he had begun so many years ago. Looking back with longing to the Nazi era of his youth, he complained that the "real values—race, nation, class, and social status" had been destroyed. As he wrote of his past, he often became nostalgic. Thoughts of his mother, the indomitable Walburga Mengele, returned to haunt him.

HEDVAH STERN:

Recently, memories of my mother have come back to haunt me. I find myself thinking, "If Mother were here, life would be better for me."

I am always thinking about her.

❧

LEAH STERN:

It is especially bad during the strawberry season. Every time I see strawberries, I remember Mother's dress when she left us. It was black, with a strawberry print on it. I think of her constantly during the strawberry season.

❧

ZVI THE SAILOR:

My mother is the person I think about the most. To be honest, I don't remember any of the other members of my family. I don't think I could even tell you the names of all my dead brothers and sisters. I can't remember what they looked like. But always, my mother is on my mind.

I think it was because she was the one who held the family together. When I was little, my father went off to war. She was the one who took care of all of us.

The first years I came to Israel, I discovered an uncle in Tel Aviv —my mother's brother. He told me there was a woman in Netanya who had been a close friend of my mother. And so, those first few years, I was constantly visiting this woman. She had a photograph of my mother as a young girl. I went all the time to talk to her, to ask her questions, to look at this photograph.

❧

LEA LORINCZI:

I was always looking for my mother—in Eastern Europe, in Israel, even in America. I would walk through the streets and look at the people. I would stare at the crowds of older women, and for a moment I would spot someone I thought looked like her. "Maybe it is her," I would think.

You always read stories about someone who survived the war, and shows up after many years.

❧

MENASHE LORINCZI:

Recently, we heard of a couple who were reunited after forty years. They met by chance in a hotel in Tiberias. Each one had thought the other was dead.

❧

ALEX DEKEL:

One day, when I was showing my American fiancée Jaffa, we passed an Arab graveyard. Immediately, I stooped down and began pulling up the weeds. She wanted to know why I was doing this—this irrational, violent tugging. "Maybe someone will do this for my mother's grave," I told her. Whenever I see an untended grave, it's hard for me not to pull up the weeds.

❧

MIRIAM MOZES:

After the war, I kept thinking, "Maybe Mother is still alive." I never knew what happened to her. There was no record of when she or my father died. I kept hoping, I kept imagining somehow that she had survived. And this went on for years and years and years.

I would see a middle-aged woman on the street in Tel Aviv and I would think, "Maybe she is my mother."

One day, as he listened to a Schubert album, Mengele remembered how fond his mother had been of that particular composition, Opus 6. As a little boy, he had surprised her as she was sobbing one day while listening to the melancholy piece, and had asked her why she was so sad. "Boy, you cannot understand that yet," his mother had replied. Listening to the Schubert album now, a dejected, tired old man, Mengele felt he knew what his mother had meant. "I understand better now, Mother," he wrote in a diary entry dated May 1978.

One weekend in February 1979, Mengele was invited to the Bosserts' cottage in Bertioga Beach. Although it would mean fun and relaxation, he kept delaying the trip. He seemed especially distracted that day, unable to complete even the simplest tasks. Finally, at three o'clock in the morning, Mengele left his São Paulo residence for the beach house.

From the moment he arrived, he turned what should have been a pleasant weekend into a nightmare. As the Bosserts would later recall both in letters to his Günzburg relatives and in interviews with the press, Mengele talked incessantly about his terrible lot in life. Although Wolfram tried to soothe him, he was not to be silenced. On and on he rambled about his life in exile—the servants who did not serve, the primitive country so unlike his beloved Germany, the boorish population with its nonexistent cultural life. These were ideas Bossert had heard him spout hundreds of times—but never perhaps with the same degree of desperation. Mengele confided that even after twenty years in Brazil, he still felt like a stranger.

The next morning, Mengele seemed calmer. But his hands shook at the breakfast table, and he spilled some coffee on himself. "This is what happens when you talk with your hands," he said jokingly to the children. He eagerly joined an outing to the mountains, and insisted on climbing unassisted. When he complained of feeling tired, his friends gently reminded him he was getting on in years, and should slow down. "I am not really that old," sixty-eight-year-old Mengele scoffed indignantly.

Bossert remembered urging him to see a doctor. But Mengele took the friendly advice as an insult. "I know you would be very relieved if I were not around any longer," he reportedly snapped. But after lunch, Mengele was in a much more cordial mood. The two men even had a good discussion.

According to the Bosserts, that afternoon, while swimming, Josef Mengele suffered a stroke. The water was stormy, and he was only able to move one arm. The Bossert son was the first one to notice Mengele was in trouble. "Come back, Uncle," he cried, "the ocean is pulling you in." The children watched in horror as their "*titio*" struggled, gasping for breath. Their father tried in vain to revive him, but it was too late. Dr. Mengele was dead.

The Bosserts tried their best to keep the drowning a secret, although the corpse attracted some attention on the small Brazilian beach. A lifeguard on the scene became suspicious at the discrepancy between the age of the man on the ID card and the much older corpse in front of him. But the couple who had shielded Mengele so skillfully when he was alive now managed to do so after he had died: Liselotte saved the day by quickly retorting their friend had suffered from an illness that "made him age remarkably."

At the beach that evening, the Bosserts say they held an elaborate ceremony in memory of the Angel of Death. There were candles and incense, prayers and hymns chanted under the stars. A wreath of red roses was placed gently on Mengele's body, while strangers who had gathered on the beach beseeched God to show mercy to the departed.

11

THE BURIAL OF THE DEAD

EVA MOZES:

When I went back to Auschwitz in January 1984, I kept looking for clues—something that would help me understand how my parents had died. I had a deep need to know what happened.

As I walked through the camp, I finally understood. I saw the path from the railroad tracks to the gas chamber, and I understood how a person could disappear from the face of the earth.

If you take a loved one who dies and bury them in a cemetery, you know you can go back and visit them. But for Miriam and me—for all the Auschwitz twins—there was no cemetery. There was only a memory of that last time we had seen our mother, our father, our sisters.

At Auschwitz, I felt at last I was at their gravesite.

In December 1979, Rolf Mengele says, he returned to Brazil to pay his last respects to the father he had hardly known. It had taken him months to save up money for the air fare. Rolf visited the cemetery in Embu where Mengele was buried under the name of Wolfgang Gerhard, whose identity he had adopted four years earlier. Rolf recalled thinking how ironic it was that his father, champion of the Aryan race, lay next to an Oriental in a small, forlorn gravesite.

Much of Rolf's trip was devoted to tidying up his father's estate. The Bosserts and Stammers, who had liquidated most of the property, were anxious to know what to do with the proceeds. The few pieces of furniture, books, and knicknacks the old Nazi had collected were hardly worth very much, and Rolf decided to let them keep the meager amount their sale had brought. In addition, he gave them his dead father's savings of about a thousand dollars. It was a small thank-you for the years both families had spent harboring the war criminal.

What Rolf cared about most, he later asserted, were the papers his father had left behind—the hundreds of pages of notebooks, diaries, calendars, and letters covered margin-to-margin with Mengele's tortuous, barely legible script. Rolf returned to Germany anxious to review his father's writings. In the quiet hours, he pored over them, in search of some special message intended just for him, even a small note addressed "to my son"—but there was none. Mengele's son also combed the papers for clues to the enigma that was his father. But that search, too, proved to be disappointing. There was far, far less of a personal nature in the papers than he had hoped.

Rolf and other family members say they decided to keep Josef's death a secret, telling only immediate relatives and a handful of intimate friends that the old Auschwitz doctor had drowned on a Brazilian beach. What were the motives of the Mengele family for keeping the secret of the grave for so many years? Explanations vary. Some have speculated that the family kept silent simply because it did not want to open the Pandora's box that did in fact open when news of Mengele's death was finally made public. Rolf Mengele himself insisted that his family merely wanted to protect the many people—Germans, Austrians, Hungarians, Italians, Brazilians, and others—who along the way had protected his father over the years.

Life went on for the rest of the Mengeles. The once-lovely Martha retired to Merano, a small town in northern Italy. She never remarried. She now leads a quiet life, reportedly traveling to Günzburg only for family gatherings. Even in her old age, there are still traces of her former beauty; she supposedly likes to wear elegant, formfitting gowns that show off her beautiful figure to its best advantage. Intensely private,

Martha declines all requests for interviews, and has never publicly discussed her life with Mengele. Her son, Karl Heinz, who is now running the Mengele factory along with his cousin Dieter, is just as silent, just as discreet about his stepfather.

Mengele's first—and perhaps only—love, Irene Schoenbein, leads a similarly solitary existence in her native Freiburg. She divorced her second husband, Alfons Hackenjos, several years ago, and he has since died. But through the years, she has continued to speak well of Josef, and has admitted to a journalist or two that she had been deeply in love with him as a young girl. If she was ever troubled by his wartime activities, she never publicly revealed her misgivings. A car accident in the 1970s left her incapacitated and nearly homebound. She keeps to herself, choosing to live in a large, secluded house with a tall gate designed to keep out all intruders.

Rolf claims he felt strangely relieved after his father's death. At last, Mengele's son could begin to lead his own life, free of the constant emotional invectives and threats his father had made over the years. With Dr. Mengele lying in a grave thousands of kilometers away in South America, Rolf could even please and honor him in small ways. Every blond, blue-eyed child Almuth bore was like a tribute to his father.

But as if on the whim of some prankish god, 1979, the year Mengele's friends claim he died, was also the year when the public's interest in finding him was most vigorously renewed. Everyone from professional Nazi-hunters and Holocaust survivors to members of Congress was suddenly expressing a passionate desire to see Mengele brought to book. Possibly as a result of the new interest in him, the Auschwitz Angel of Death was now spotted regularly in bustling Latin cities and dense jungles south of the Amazon. He was handsome and dashing as ever, witnesses said, as he shuttled around the triangle border area of Paraguay, Argentina, and Brazil.

For years, Simon Wiesenthal and others had been pointing the finger at Paraguay. At around this time, Wiesenthal was joined by a new and dynamic breed of Nazi hunters, led by the husband-and-wife team of Serge and Beate Klarsfeld. Serge, a French Jew, and Beate, a German Gentile, didn't just stop at condemning the South American dictator who harbored Nazis. They traveled to their countries and waged fiery demonstrations. The Klarsfelds, like Wiesenthal, insisted the regime of General Alfredo Stroessner was harboring the war criminal. As a result of the clamor, Congress joined the fray, urging the

State Department to apply pressure on the Paraguayan government. Stroessner and his deputies had always maintained they knew nothing about Mengele's whereabouts. But the U.S. ambassador to Paraguay, Robert White, was inclined not to believe a regime that was one of the world's worst human-rights violators. Convinced that the government that had once granted the Auschwitz doctor citizenship was protecting him still, White was certain that Mengele was living somewhere in the vicinity of Asunción. He took to wandering alone through the German establishments of the capital in search of information, and sent regular cables to Washington detailing the progress of his "hunt."

In a surprise move that year, the Paraguayan government caved in to pressure and revoked Mengele's citizenship—some two decades after granting it to him. It was a gesture clearly intended to appease critics of the Stroessner regime, particularly those in Washington.

That same year, one of the strangest reports ever concerning Mengele arrived by way of the State Department's secret channels. The embassy in Asunción had learned, ostensibly from good sources, that Dr. Mengele would be flying to Miami, Florida, on a Delta Airlines flight. Although information was sketchy, reservations were found to have been made under his name. The bizarre report was taken seriously at the highest levels of the State and Justice Departments, and a major effort was launched to arrest Mengele when he landed. Classified cables flowed back and forth between Washington and Asunción. The matter even reached the desk of then-Secretary of State Cyrus Vance. It was Vance who gave the orders to arrest Mengele when he arrived in Miami.

But, of course, Josef Mengele never boarded the Delta flight. FBI agents who were poised to nab him at the Miami airport returned, empty-handed and disappointed, to their headquarters. They subsequently filed a report dismissing the incident as a probable hoax. But no one was ever able to explain the mysterious reservations made under Mengele's name.

"If I could get this man, then my soul would finally be at peace," Simon Wiesenthal liked to tell friends and colleagues. He continued to tantalize the world with fantastic tales of the Auschwitz doctor. One day, Mengele would be spotted in a Mennonite community, deep in the heart of Paraguay. The next, he was in Bolivia, fraternizing with Klaus Barbie; the Angel of Death and the Butcher of Lyons were rumored to be on excellent terms. A priest spotted him on the Brazilian border. A Holocaust survivor swore she had seen Mengele shopping in her store in downtown Asunción. Uruguay had asked him to train its police force. Stroessner had summoned him for expert medical advice. The CIA had reports he was involved in the international narcotics trade. He was rumored to be best friends with another missing Nazi, Martin Bormann. He was dying of cancer. He had just had a face-lift. He looked haggard and worn out from his years on the run. Mengele was seen here and there, everywhere and nowhere. The more Wiesenthal floated reports about Mengele's supposed whereabouts, the more elusive the Angel of Death seemed to be.

The 1980s brought a resurgence of interest in the Holocaust. In 1983, the first national conference for survivors in Washington attracted some fifteen thousand victims of Hitler's death camps. They gathered to search for loved ones and friends and relive the horror. Many spoke for the first time about their experiences. They crowded around specially installed computer banks crammed with names of survivors hoping to find a long-lost relative. The historic three-day reunion prompted an appearance by President Ronald Reagan, who made a moving speech about America's responsibility to remember the Nazi slaughter of the Jews.

Attending the gathering was Mengele's child-victim Alex Dekel, who was in Washington to lobby members of Congress about Mengele. Obsessed with finding the war criminal, Dekel wanted the United States Congress to pressure Paraguay to turn over Mengele. But within weeks of the gathering, Alex died—and with him, it seemed, any resolve to capture the missing death-camp doctor.

But the following year, Eva Mozes, a surviving twin who had also been in Washington, earnestly took up Alex's cause. She and her sister, Miriam, decided it was time the world should learn the story of the Auschwitz children. From her home in Terre Haute, Indiana, Eva sent letters to five hundred prominent American journalists and newspapers, urging them to write about the missing Auschwitz doctor and his child-victims, the twins. For months, she waited for a reply to her anxious requests. One day, her photocopied letter reached the desk of nationally syndicated columnist Jack Anderson. Anderson received thousands of letters a week, and often turned them over to his team of investigative reporters. Struck by the poignancy of Eva's letter, Anderson associate Lucette Lagnado decided to contact the Indiana housewife. In the course of several emotional phone conversations,

Eva told Lagnado of her childhood experiences as a surgical guinea pig, the brutal loss of her parents and sisters in the gas chambers of Auschwitz, and life with Dr. Mengele.

EVA MOZES:

Over the years, I had had so many nightmares about Auschwitz. I was always thinking about Mengele, about the camp, about the other twins.

I had read many of the books about Auschwitz, but to my great sorrow they never mentioned Mengele and the twins, except in passing.

I had a very personal need to search for the twins. I wanted to understand what had been done to Miriam and me—to know why she was ill so often, why I had so many problems.

I thought that maybe if I could locate the twins, we could sit down and piece together what had happened to us in the concentration camp.

And then one day, it occurred to me that there was one person who knew exactly what was done to us—Dr. Mengele. I found out that Mengele had been free since the end of the war.

I thought, "That is impossible."

In Israel, Eva's sister, Miriam, was equally intent on locating other twins. With the help of a relative who worked at *Ma'Ariv*, one of Israel's leading dailies, Miriam placed a small ad asking any child survivors of Mengele's experiments to contact her. Within days, she was flooded with calls and letters from twins longing to see each other again. A reunion was quickly arranged.

HEDVAH STERN:

I had been waiting for this moment—for someone to bring up what had happened. The wound was there, and it reopened.

❧

LEAN STERN:

I kept crying—shouting and crying. "Why, why, why didn't anyone do this before?" I blamed the world for the fact that no one had ever taken notice of us.

❧

MOSHE OFFER:

There were so many of us still around—some had been hurt less, some were hurt more. I went to the reunion thinking that all the twins who underwent experiments at Mengele's hands were suffering. And their children were suffering. And they would suffer for generations to come.

We wanted to know what had been done to us. We felt the world should research the injections and surgeries we had undergone.

❧

ZVI THE SAILOR:

I walked in thinking, "I should know half the people in this room." A pair of twins immediately came up to me, and said hello.

They told me, "How could we forget you—you and your brother were always fighting." And one of them remembered not only my name, but my number—and even my twin's number. I checked my arm, and sure enough, it was the same as he remembered.

❧

JUDITH YAGUDAH:

As the years passed, I had tried not to think about Auschwitz at all. I tried to push it back and live in the present.

On the other hand, whenever a book about the Holocaust came out, I ran to get it.

Auschwitz was in me—like a heavy parcel I had to drag around all the time.

After the reunion, I felt very sad. I had a very hard week. I thought about Ruthie a lot. I talked with my mother about her. I told my children stories about her. I lit a candle for her.

Buoyed by the discovery of nearly a hundred twin survivors, Eva and Miriam decided to form CANDLES, an organization whose sole purpose would be to publicize the plight of Mengele's child victims. *Parade* magazine's publication in September 1984 of Anderson's feature story on the Auschwitz twins helped focus the spotlight on CANDLES. Thousands of readers around the United States sent contributions, and other journalists also were spurred to begin writing about the unusual group. The attention prompted Eva and Miriam to plan the most dramatic project ever undertaken by a group of survivors: a return to the death camp.

In January 1985, CANDLES traveled to Auschwitz to commemorate the fortieth anniversary of the camp's liberation from the Nazis. Led by the Mozes sisters, twins from Israel and America braved once again the frigid Polish winter and their own demons for a walk through Auschwitz. Journalists from around the world descended on it to hear the forty-year-old story of the twins' ordeal under the abominable Dr. Mengele. With tears and prayers, the twins toured the relics of the camp, anxious now to remember something of their terrible childhood past, the past they'd spent all their adult years trying to forget.

MIRIAM MOZES:

It was very terrible to go back. I felt like I was going to a funeral. Auschwitz had been the last place I had seen my mother, my father, my older sisters.

❧

EVA MOZES:

We walked to the spot where the twins' barracks had stood. There was only the foundation left: We learned that the Poles had destroyed it and used it for firewood.

But everything around the twins' barracks was just as I remembered it.

There was the watchtower! And here were the brick barracks that were used to warehouse the dead—just the same as when I had left it, forty years before.

❧

MENASHE LORINCZI:

I decided to go to the crematorium by myself. I left the group and walked inside. I started to read the psalms of David.

I never cry—often, I have wanted to cry, but I could not. But there, in the crematorium, I found myself crying. I prayed, and then I cried.

❧

MIRIAM MOZES:

When someone dies in the Jewish religion, it is very important to bury them quickly. You are then expected to "sit shivah*"—to mourn them. But the mourning period is strictly limited.*

All these years, I never buried my parents. There was never any funeral. I never did anything in their memory, not even to say a memorial prayer.

Instead, for years, I kept on mourning their loss.

Oddly enough, I felt strangely free at Auschwitz*. At last, I had found my mother's resting place. I could speak to my mother there. It was the only place in the world I felt close to her. Our liberation had happened in 1945, but I felt personally liberated for the first time in 1985. I felt I could stop looking for my mother. I knew I had found her resting place.*

Finally, the twins retraced the Death March in which thousands of prisoners had trudged through the snow, just days before Soviet troops arrived. And as they marched, Mengele's children began to sing—loud, boisterous Hebrew songs.

EVA MOZES:

One of the reporters asked me, "Why are you singing?" I told him that I felt strangely upbeat. I had never walked through these grounds as a free human being. Forty years before, I had been a skinny kid, half-dead, an orphan. Then, we could have died like flies, and it wouldn't have mattered. Now here I was, surrounded by the world press.

Since the Holocaust, I had always felt that if only the world knew our story, it would care. At Auschwitz*, surrounded by all those reporters, I felt, People do care.*

The whole time I was at Auschwitz*, I had this wonderful feeling my mother was there also, listening to me, watching over me. If I could, I would go back to* Auschwitz *every year.*

After Auschwitz, the twins returned to Israel and held a mock trial of Dr. Mengele at Yad Vashem, Israel's monument to the six million

Jewish Holocaust victims. Nearly thirty twins and dwarfs testified before a six-man tribunal of world dignitaries. General Telford Taylor, whose team of Nuremberg prosecutors had let Mengele slip away, who had dismissed Mengele as dead, was on hand for the trial. And so was Gideon Hausner, who had successfully prosecuted Adolf Eichmann. Simon Wiesenthal, who had kept the memory of Mengele and so many other war criminals alive when the world would rather have forgotten, occupied a place of honor on the podium.

Shyly, one by one, Mengele's twins stepped up and, under the glare of the TV lights and cameras, told the world of their ordeal. Moshe Offer addressed the tribunal from behind a curtain. In a halting voice, he spoke of his brother Tibi's ordeal under Dr. Mengele, the successive surgeries he had undergone, culminating in his barbaric castration and death. Zvi Spiegel, who had been known only as Twins' Father, recounted his efforts to care for and protect the boy twins even as he feared Mengele's wrath. When Spiegel's presentation was over, all the boys who had been under his ministering rose to greet their Twins' Father. A former nurse at Auschwitz recalled watching Mengele "sew" two twins together in an effort to make them Siamese. A pair of female dwarfs—also twins—wept as they remembered being forced to perform naked in front of Mengele; before the war, they had been circus performers. How delighted Mengele had seemed when they were first brought to him! He had chuckled, "Now, I have work for twenty years."

In the audience, dozens of twins from Israel, the United States, and around the world listened to the testimonies silently, intently, often tearfully, As for the Mengele family in Germany, if they followed the "trial" of the man they had successfully shielded all his life, they said not a word. Was Rolf Mengele shocked by Moshe Offer's story of Mengele's slaughter of his twin brother? Did Karl Heinz, in his office at the Mengele farm-equipment factory in Günzburg, wince at the accounts of his stepfather's mutilations and grotesque experiments? Did Dieter Mengele feel any shame? Did Irene? Did Martha? No available evidence suggests that they did. Indeed, none of the Mengeles, not his son, not his nephews, not his wives, have ever expressed any remorse or any wish to atone for the massive suffering inflicted by their notorious relative. Even Rolf, the most public of the Mengeles and supposedly the most contrite, has been ambiguous and contradictory in his statements.

At the close of the historic three-day proceedings, the tribunal put out a statement saying, "There exists a body of evidence justifying the committal for trial of the SS Haupsturmführer Josef Mengele for war crimes and crimes against humanity." What Taylor and others had been unable to do at Nuremberg, they were attempting to do now.

CANDLES' pilgrimage to Auschwitz and trial of Mengele galvanized the moribund forty-year hunt for the war criminal. Until 1984, no government had cared enough about Mengele to launch a sustained effort to find him. But now, inspired by the twins' stories of life in the Mengele barracks, governments and individual Nazi-hunters competed to find him. Large rewards were proffered for his capture. The West German government, which had done little to pursue Mengele since the 1964 Frankfurt trials, now offered $300,000 for any clues that might lead to him. In California, the Simon Wiesenthal Center posted a reward of $1 million, while in Washington the *Washington Times* newspaper promised another million dollars. In May, the Israeli government also put up $1 million. By the spring of 1985, there was a price of nearly $4 million on Mengele's head.

Enticed by the promise of millions, fortune-hunters took off for the jungles of South America to search for Mengele. Camera crews and squads of television correspondents became a commonplace sight in the most remote parts of Paraguay and Brazil.

The West German and Israeli governments—traditionally distrustful of each other—gingerly began to cooperate to find Mengele. As neither wanted to share their intelligence with the other, the United States was forced to act as an intermediary. Yet, when officials from the U.S. Department of Justice actually examined the "intelligence" available in the files of the Germans and Israelis, as well as whatever material the CIA possessed, they found it amounted to very little. There had not been a single reliable sighting of Mengele in over twenty years. There was no recent photograph known for certain to be of him. Although he was said to be in or around Paraguay, the experts who gathered to discuss the case agreed he could be anywhere. Four decades of sporadic hunts by various individuals had resulted in only a stream of rumors and dead ends. He had become a mythical creature. Fantastic tales—and few facts—about Dr. Mengele abounded.

The Justice Department's Office of Special Investigations (OSI), which was conducting the search for the U.S. government, found itself deluged with an inordinate number of bizarre leads. Mengele was spotted in Las Vegas, in Westchester, and even playing in a marimba band in California. At one point, investigators were again directed to

Miami Beach, to the Fontainebleau Hotel, where a clerk swore a man by the name of Mengele had checked in. When OSI officials arrived at the elegant resort, extremely popular with observant Jews, they found a Mr. Mengele was indeed registered. They hurried up to his room, banged on the door, and were greeted by Dieter Mengele—Lolo's flustered son.

The driving force behind the Justice Department's inquiry was the memory of the Klaus Barbie investigation, which found that the United States had long been secretly involved with the infamous Butcher of Lyons. Because Mengele had remained at large for so many years and had never been tried, there was a suspicion the United States had helped him, too. But files at the National Archives—which had contained the damning evidence linking Barbie to American intelligence—were sketchy when it came to Mengele. The most intriguing document suggested that Mengele had been interrogated at a prisoner-of-war camp (something he alluded to in the autobiographical novel), then let go. But there was nothing to suggest an active involvement between the Auschwitz doctor and the American postwar government.

Because of the scarcity of solid clues, the German, U.S., and Israeli team of investigators decided the surest way to find Mengele would be through his family. For years, Hans Sedlmeier, the trusted Mengele family assistant, had been suspected of maintaining links with the war criminal. Sedlmeier, who was now living quietly in Günzburg, had openly admitted helping Mengele years earlier in an interview with American journalist Flora Rheta Schreiber published by *The New York Times* syndicate in the early 1970s. At the time, Sedlmeier's admission went unnoticed; no one seemed to care.

A surprise search by German authorities of Sedlmeier's house in the spring of 1985 uncovered dozens of letters ostensibly from Mengele. Unbeknown to Sedlmeier, his wife had kept copies of letters from the war criminal over the years. Searchers also found a notebook with the names and Brazilian addresses of both the Bosserts and the Stammers. Orders were immediately given to find and interview them. In the course of an interrogation, Liselotte Bossert quickly volunteered that Mengele was dead, and directed searchers to his gravesite.

In June 1985, an international team of forensic experts gathered around the unkempt gravesite in Embu, Brazil, marking the remains of Wolfgang Gerhard. As hundreds of reporters and cameramen watched intently, the experts broke open the casket lid to reveal a grisly heap of bones. The scientists pulled out a pile of bones, some teeth, a few clumps of hair, and a pair of faded, rotting trousers. The scientists basked in the unprecedented media spotlight. Pressed to come up with instant answers to satisfy the curiosity of the journalists around them, they made rapid determinations based on superficial examinations and incomplete data. The scientists competed with each other for the attention of the television cameras. Within three days, they had declared with "scientific certainty" the remains were those of the Angel of Death, Dr. Josef Mengele.

Soon after, Wolfram and Liselotte Bossert and Gitta Stammer emerged for the first time to talk to the press about their years aiding the Auschwitz doctor. The Bosserts showed several photographs they had kept of the war criminal, and even held up his beloved Burberry raincoat. Former neighbors of Mengele in São Paulo also surfaced to talk almost fondly about the private, elegant old man who kept to himself as he took long walks through the dilapidated neighborhood.

Draping herself in the soft white shawl Mengele had given her years before, Elsa Gulpian wistfully recalled her years with the Auschwitz doctor. Elsa told story after story about the gentle old man she had fallen in love with years before, and who had refused to marry her. Meanwhile, Rolf Mengele also came forward to publicly assert that his father had indeed died in 1979. Rolf released a wealth of papers, diaries, notebooks, and photographs to the German magazine *Bunte* for publication. There are suspicions that Mengele's son sought to profit from the publicity surrounding his father's death. It has been claimed that he became involved in a book deal, and he even sought a Hollywood contract to make a movie about his father.

The U.S. team of forensic pathologists was summoned to testify before Congress about the Embu skeleton. The scientists insisted the bones were indeed Mengele's. Under oath, they swore there were no inconsistencies between the skeletal remains found in Embu and what was known about Mengele from his military and medical records. But an attorney at the World Jewish Congress, now the OSI, raised questions about the doctors' findings. All of them knew, after all, about Mengele's teenage bout with osteomylitis, which appeared in his war records. The virulent bone disease ought to have turned up on the skeleton, and the scientists' inability to find even a trace of the osteomylitis certainly represented an "inconsistency," one, moreover, the forensic scientists ought to have revealed to Congress.

As, one by one, Nazi-hunters like Wiesenthal and Serge and Beate Klarsfeld declared themselves satisfied with the findings, the German, American, and Israeli governments also called off their hunts. In Washington, then–attorney general Edwin Meese announced that the U.S. search for Dr. Mengele was over.

Behind the scenes, however, investigators of all three governments were at first reluctant to close their Mengele files. The Mengele family had pledged to turn over medical records and X rays that could help identify the skeleton in Embu, but never did so. Nor was Germany able to turn up any X rays from Mengele's youth that would have enabled scientists to show conclusively the remains were those of the Auschwitz doctor. In Washington, Frankfurt, and Jerusalem, experts continued to ponder the troubling question of the osteomylitis. Forensic scientists were asked again and again to reexamine the skeleton for signs of the disease, which, by all accounts, should have left marks, even years later.

Anxious to settle the question once and for all, in 1986, the Justice Department even hired an expert from the Smithsonian Institute to fly to Brazil and reexamine the skeleton for traces of the osteomylitis. But he, too, could not find definite proof of the illness.

With the exception of the dental forensic scientist, members of the U.S. team of experts never did submit a final report on their findings, a comprehensive study that would have laid to rest any lingering questions about the authenticity of the skeleton. Despite their rush to proclaim in front of TV cameras that the Embu skeleton was Mengele, the pathologists evidently would not put their findings on paper. Only with the discovery, nearly a year later, of dental X rays in Brazil believed to be Mengele's, did one team member, forensic odontologist Dr. Lowell Levine, submit his own report to the Justice Department. It was never made public because the case was never closed.

Meanwhile, a fickle public and press quickly forgot that no final report was ever issued on the Mengele investigation. Despite the vast amounts of money, time, and staffing devoted to the hunt, it was never possible to review the body of information the Justice Department had gathered. Although privately, officials confidently insisted that Mengele was dead, they did not say so publicly. One problem hampering U.S. officials was an agreement made at the start of the 1985 hunt between Frankfurt, Tel Aviv, and Washington: that all three governments would publicize their Mengele reports at the same time. But in Israel, one man was reluctant to officially close the case, and because of his adamance, he succeeded in preventing Germany and the United States from doing so.

After the discovery of the bones, veteran Nazi-hunter Menachem Russek had quietly undertaken an investigation of his own. A colonel in the Israeli police—and an Auschwitz survivor—Russek was profoundly disturbed by the speed with which other countries wanted to close their Mengele files. Russek flew to Brazil and Germany to interview individuals connected with the case. Russek also spent months going over the various diaries and notebooks said to have belonged to Mengele. Russek concluded the body was part of an elaborate hoax designed to throw the Western world off the scent.

He prepared a scathing forty-page report, in which he wrote that "it is not yet time to 'bury' the Angel of Death, Dr. Josef Mengele." Russek thought the six-year-old "death by drowning" story on a Brazilian beach much too tidy. Although other top officials in the Israeli government believed the skeleton in Embu was that of Mengele, Russek kept stubbornly insisting the case should remain open. This veteran of Nazi hunts was mocked behind the scenes as an old sentimentalist who could not bring himself to admit Mengele was dead. The Israeli government would not even allow him to release his report. He was later edged out of his job, and pressured into retirement.

Mengele's twins do not for a moment believe their nemesis has died. They are sure the Auschwitz doctor has succeeded in tricking the world yet another time. Bewildered and betrayed, they cannot understand why the hunt for him has been abandoned.

Mengele, who was an enigma in his youth, a riddle at Auschwitz, and elusive in his life after the war, remains a mystery even after his supposed death.

MOSHE OFFER:

Dr. Mengele was a very shrewd, a very clever man. I have a feeling he will never be caught. A man who gives children sweets, then terrible injections—do you really believe such a man would let himself be caught?

In 1987, the German government awarded the Auschwitz twins thousands of dollars each in restitution—a small, belated attempt to make up for the pain they had endured, an effort to erase the past. Of course, none of Mengele's children will ever be able to forget what happened. Images of the Auschwitz death doctor are still as vivid and terrifying as decades earlier. But thanks to Germany's action, as well as the attention afforded to their plight during the mock trial and manhunt, a few of them have begun to find it in their hearts to forgive. With forgiveness comes a faint sense of optimism, a glimmer of the inner peace that has eluded Mengele's twins for so long.

VERA GROSSMAN:

In the spring of 1987, I was invited by a German pastor of a church in Frankfurt to come and give a lecture. I was to go as a representative of the Mengele twins, and speak about "building bridges" between Israel and Germany, between the victims and the perpetrators.

I agreed, excited about the opportunity of speaking to Germans about my experiences in the Holocaust.

At the church, I found I was one of a panel of speakers—nearly all of them Germans. I gave my talk in English, and was praised and applauded.

Afterwards, I went to a reception in my honor in the church hall. As I stood greeting people, I saw two middle-aged women come up to me. Very politely, they asked me if I was "Fraülein Grossman." I said yes. It was then that they told me they were the daughters of Professor Otmar von Verschuer.

I felt faint—I felt my legs go under. I started praying that I would not collapse—I prayed to my Jewish God inside this church that I would keep my composure.

We began to talk. We went to sit on a bench in the church. They wanted to know what their father had done. I told them what I knew: that Verschuer had been both Mengele's superior and his mentor. We spoke for nearly two hours.

Then they told me their story. They told me they were little girls —ten and twelve years old—when Mengele would come to their house to visit their father. They remembered how the two would talk for hours about music and science.

Mengele often brought candy and toys for them. They loved him! They even called him "Uncle Mengele."

Afterword

CHILDREN OF THE FLAMES—

The Roll Call

On January 27, 1945, the Russian Army marched into Auschwitz. The liberators entered a cold and forlorn barracks, and there they found approximately 160 of Mengele's twins crouched within. Later, they discovered that another thirty or forty additional twins had also managed to survive. Still later, Israeli and Polish scholars learned that a total of three thousand twins had passed through Mengele's experimental laboratories. Their survival rate had been less than 10 percent. Liberation had come in the wintry chill of Polish soil, but the one simple, terrible historical fact was that more than twenty-eight hundred of the twins would not be coming back.

Today, the surviving twins are spread out across four continents and at least ten countries. These wounded adults have never quite stopped thinking of themselves as Mengele's child guinea pigs. And while their spiritual bond is deathless and infinite, they have scattered into the four winds. Today, they can be found everywhere from Brooklyn, New York, to Sydney, Australia; from Skokie, Illinois, to Zurich, Switzerland; from Jerusalem to Budapest; from Tel Aviv to Brussels.

But wherever they are, wherever they go, the Children of the Flames will always awaken to the noise of Mengele's footsteps, to the sounds of the Auschwitz Roll Call:

ZVI THE SAILOR Zvi Klein is still sailing, though much closer to home. The most restless of Mengele's twins now works for a local shipping company in his hometown of Ashdod, Israel. His exotic voyages to distant lands have ended, but he still manages to be away at sea three or four days a week. The strapping Zvi, who is sixty years old, insists he has the stamina of far younger sailors. He lives with his wife, who has a son from her first marriage whom Zvi helped raise and loves as his own. The vagabond of yesteryear is now a grandfather of four. His only sadness comes from his continued estrangement from his twin brother.

THE STERN SISTERS At sixty Hedvah and Leah are still inseparable. Neither has lost the optimism and sense of humor that enabled them to overcome Auschwitz. During the war, the Stern sisters were always delighted when the Red Cross trucks that transported them to Mengele's laboratory could not come, and they were made to walk through the camp. Even though a stroll through Auschwitz could hardly be scenic or enjoyable, the twins were happy for the exercise and glad to be outdoors.

And today, the sisters who loved fresh air, even Auschwitz air, are living out their lives on the *Moshav*, the communal farm they helped found in Ashdod. Both continue to work in the fields. Both are married, both have several children and grandchildren, and both are hoping, any day now, to become great-grandmothers. Occasionally, they pause and think wistfully back to their days growing up in Eastern Europe, when, together with the mother they adored and lost, they had hoped to open up a seamstress shop—The Stern Sisters.

MOSHE OFFER Moshe has a high-level job as a technician for Israeli television. But the images he sometimes glimpses in the course of his work have nothing to do with daily program fare. In the middle of the most innocuous daytime soap opera, "Miki" sees his twin brother, Tibi, pale and sickly and entreating, the way he looked after returning from a session in Mengele's laboratory. The images on the screen become mangled and confused.

Other days, during a news show, perhaps, or the screening of an old movie, Offer is transported back to his parents' estate in Transylvania, in those joyful years before the war, when he and Tibi were spoiled and lavished with gifts. He sees Tibi smiling and laughing in the horse-drawn carriage reserved just for the twins. Together, he and Tibi secretly take sips of the sweet, syrupy liqueur their father manufactures. The flickering images on the screen become clear again.

At fifty-eight, the little boy who survived the loss of his parents, brothers, sisters, and finally of his twin, and who endured several nervous breakdowns, has rebuilt his life. Although he dearly loves his Iraqi-born wife and their many pretty dark-eyed daughters, the pride and joy of his life is his lone son Shai, who reminds him of Tibi.

TWINS' FATHER The oldest of Mengele's twins, Zvi Spiegel is seventy-five now. The man who led an army of twins out of Auschwitz has slowed down. He retired several years ago from his job as an accountant and spends much of his time in his Tel Aviv home, enjoying the company of his wife, his two children, and five grandchildren. Occasionally, he will get calls or letters from his other children—the twins. That's when he'll drop everything and become *Zwilingefater* once more, listening to them with a sympathetic ear and attending to their every need.

MAGDA SPIEGEL Zvi's sister lives in Haifa with her husband, Nahman. Magda, whose six-year-old boy was sent to the crematorium by Mengele, is a mother of two and a grandmother of four. She is especially proud of her son, who became a doctor and has received many honors for his scientific research.

But when Magda is alone in her Haifa apartment, she furtively pulls out the old photographs of the blond, angelic little boy she lost. And then she remembers the masses of tired Jews trudging down the selection line. And she relives that fateful moment when the handsome Dr. Mengele stepped up to ask her the question that both saved and damned her: Was she a twin? For not a day passes that Magda, guilt-ridden and inconsolable, does not think of her dead son.

EVA AND MIRIAM MOZES Eva Mozes, now Eva Kor, lives in Terre Haute, Indiana, where she is a successful real estate agent. She is married to a pharmacist and has two children, a boy and a girl. Eva runs CANDLES' worldwide headquarters from a cramped spare room in her Indiana home. Her sister, Miriam, resides with her husband in a small house in picturesque Ashkelon, an ancient seaside resort in Israel where the biblical Golden Calf was recently unearthed by archaeologists. Miriam has three daughters and five grandchildren. Though she has been quite ill of late, Miriam continues to work as a nurse in a nearby hospital. Because of her calm manner and cool professionalism, she has been promoted to head nurse, in charge of the heart-and-lung clinic.

But Miriam, who was diagnosed as having a rare form of cancer, needed some nursing care of her own recently. Who best to call on but Eva, who promptly flew in from Indiana to take care of her twin. And when the doctors declared that Miriam could only be saved if she received a new kidney, Eva unhesitatingly volunteered her own. Dozens of Mengele's twins dropped by to visit them before the surgery. The Mozes sisters, alive and well with one kidney each, just turned fifty-five.

JUDITH YAGUDAH After years of financial struggle, Judith and her husband are at last enjoying some measure of success and stability, and savoring the rewards of many years of labor. They have moved to a lovely mansion in Savyan, outside of Tel Aviv—one of Israel's most fashionable and elegant suburbs. Judith recently suffered the loss of her mother, Rosie, and was absolutely devastated. She now thinks a great deal about those years with her and aches at how much she suffered over the death of Ruthie. Judith's great solace comes from puttering around the garden of her villa. It conjures up memories of the lost garden of her childhood in Cluj, where she and Ruthie retreated daily to play under the watchful eyes of their mother. She remembers how fond Ruthie was of the simple act of gathering up stray leaves, and how she would hand them over solemnly, as if they were a great and precious gift to her twin.

PETER SOMOGYI Peter lives in Pleasantville, New York, in affluent Westchester County, where he has opened a successful appliance business. He and his wife have two children, a boy and a girl. He is fifty-seven years old.

LEA AND MENASHE LORINCZI Lea Lorinczi, now Lea Gluck, runs two flourishing ladies' garment stores in Manhattan's Lower East Side. She and her husband are devoutly religious Hassidic Jews, and live in Williamsburg, Brooklyn. They have two children and several grandchildren. Menashe, her twin, lives with his wife Yaffah in Netanya, where they spend most of their time caring for their three children and six grandchildren. The Lorinczi twins are fifty-six years old.

VERA GROSSMAN Vera is the gypsy among Mengele's children. At fifty-two, she can be found wandering around Europe, the United States, and Israel. A popular figure on the lecture circuit, she is also devoted to her daughter, Irit, her son, and her three grandchildren. And she makes frequent trips to Haifa to look after her twin, Olga.

OLGA GROSSMAN Olga is slowly finding in middle age the peace of mind that eluded her in her youth. She has recovered sufficiently from her depression to stop seeing Dr. Stern and resume a normal life. She is still happily married to her army colonel, and together they live in a sun-drenched apartment high on the hills of Haifa. They have two children. Olga is excited about the prospect of becoming a grandmother; she hopes it will be less painful than being a mother.

EVA KUPAS Eva is a housewife and mother, living in Tel Aviv. She is intensely private about her life, and discloses few personal details.

VERA BLAU Vera, fifty-six, has become an artist in Tel Aviv. Vera's art instructor is persuaded that his lively, sensitive student could add a great deal to the existing body of art on the Holocaust. He urges her to paint scenes from the concentration camp. But this veteran of Auschwitz, Theresienstadt, and Bergen-Belsen has no desire to recapture the past: She looks to the present for subjects she can paint. Her canvasses are bright and colorful, depicting joyous scenes of everyday life. Her paintings are sunny and cheerful, and that is perhaps not as improbable as it seems. They were painted, after all, by the grown-up little girl who still believes, in spite of everything she has seen and heard, that Dr. Josef Mengele loved children.

DR. JOSEF MENGELE In his life, the Auschwitz doctor wandered far afield from his native Germany. He journeyed to Poland, to Russia, to Yugoslavia, to Austria, to Italy, to Argentina, to Switzerland, to Paraguay, to Uruguay, and finally to Brazil. His wanderings have not ceased. His alleged skeletal remains were at last returned to the Fatherland in 1989. But still the bones keep moving. And in 1990, forty-five years after the Liberation, the Allies finally got their hands on the Angel of Death. After a lifetime on the move, he has made another voyage, this time to England. The bones alleged to be his were recently shipped across the Channel to the University of Leicester, for forensic and pathological testing performed by yet another medical expert. Somehow, it is hard to imagine that the story will end there in the British Isles.

As with everything else concerning Dr. Mengele, the analysis is proving difficult, the results, elusive.

Acknowledgments

The research and writing of *Children of the Flames* took me over six years, and required the cooperation of hundreds of Mengele experts, scholars, victims, and "friends" (yes, there were a few) from Washington to Warsaw, Asunción to Auschwitz, Günzburg to Jerusalem.

Children of the Flames dealt with one of the most evil doctors the world has ever known, but it was inspired by a very kindly and gifted physician, who has treated me since I was a very young girl. His name is Doctor Burton Lee, formerly of Memorial Sloan-Kettering, and now with the White House, where his star patient is none other than the President of the United States. Amazingly, he still continues to see me, and I am deeply honored by the privilege. Dr. Lee provided an extraordinary amount of encouragement through the years, and always made me feel hopeful—perhaps, the greatest gift a doctor can bestow. I would also like to thank another physician, Dr. Jerome Breslaw, of the Beth Israel Medical Center, for also having worked miracles.

I am deeply indebted to Andree Pages for her gifted and inspired editing. She devoted an extraordinary amount of time fine-tuning this book, and provided a great deal of encouragement and reassurance. She proved to be a brilliant editor and friend.

I also would like to acknowledge the late Dr. Werner Lowenstein for his investigations of Mengele which were so helpful in my general research.

This book would not have been possible without the support of my former boss, syndicated columnist Jack Anderson, and his ex-partner, journalist Leslie H. Whitten, Jr. One year before the hunt for Mengele came to dominate the international media, Jack had the instinct to assign me to the story and the search. But it was Les who gave me the faith and courage to write this manuscript, and would never let me falter. I shall always be grateful to him.

I am thankful to the U.S. Justice Department's Office of Special Investigations, whose officials shared with me the fruits of their labors. Neal Sher, OSI's director, was extremely supportive, and allowed his staff to help me in every way. David Marwell, his chief historian—and my personal friend—provided me with constant and unlimited and unwavering support for this undertaking over the six-year period. Marwell, who was in charge of the Mengele investigation for OSI, was there day and night to respond to my most detailed queries on Auschwitz, the Holocaust, World War II, and, of course, Dr. Mengele. Now director of the Berlin Document Center, he still fields my overseas calls and my continuing million and one questions about Mengele. David's wife, Judy, edited large sections of this manuscript, and was instrumental in devising a way to merge the text with the "Greek Chorus" of the twins. I am deeply grateful to both.

My most delightful surprise in writing *Children of the Flames* came about while doing research in Germany. Countless professors, students, historians, and journalists went out of their way to assist me with a task that could not have been easy or pleasant. In the course of my encounters with these men and women, all living examples of the "new" Germany, I came to fervently believe in the possibility of change and the notion of redemption.

I wish to thank the editors of *Bunte* and *Stern* magazines for having provided me with a wealth of original material on Mengele they had acquired from the family. *Stern* was kind enough to hand over to me literally thousands of pages of Mengele diaries and notebooks, which had never been reviewed by any American writer. Similarly, *Bunte's* gifted editor Norbert Sacowski, whose magazine was the first to publish the Mengele documents, was personally most helpful and encouraging. *Bunte's* brilliant five-part series on Mengele was an invaluable reference source in the face of the Mengele family's steadfast refusal to release the papers to the general public.

A number of German scholars and journalists also provided critical assistance. Chief among them was Dr. Zdenek Zofka, self-appointed historian of Günzburg, who responded to my transatlantic queries over a six-year period, and was unfailingly pleasant and polite in answering my most arcane questions about Mengele's Bavarian youth and upbringing. I am also indebted to Dr. Klaus-Dieter Thomann, whose seminal work on Nazi scientists, particularly his research on Mengele's mentor, Dr. Otmar von Verschuer, proved invaluable.

Two young American women of German background helped me to translate the thousands of pages of German-language documents that were used to write *Children of the Flames*. They are Charlotte Hebebrandt, now with the U.S. Holocaust Museum in Washington, and Elizabeth Bergmann. I thank them for their patience in deciphering Mengele's often-illegible handwriting, and poring through the many diaries and notebooks. Maria Guitard also assisted in the translations of several Mengele letters and notebooks.

Several international archives and museums provided me with material that appears throughout the pages of this book, including the U.S. National Archives, the Berlin Document Center, and Poland's Auschwitz Museum. The Polish government was most helpful in handing over the files of its now-defunct Military Commission for the Investigation of War Crimes for my research.

I am especially thankful to Dr. Robert Wolfe, director, Captured German Records, at the U.S. National Archives, for helping me to retrieve all sorts of extraordinary documents relating to Dr. Mengele. His late colleague Dr. John Mendelsohn was also very helpful to me, and I mourn the fact that he is not around to see the book come to light.

In Paris, Natzi-hunters Serge and Beate Klarsfield were very helpful in providing early tips and leads that helped me put together the life story of Josef Mengele.

In Vienna, Dr. Ella Lingens, a survivor of Auschwitz who worked alongside Mengele as a doctor, offered invaluable insights into his character and personality.

In Frankfurt, Hans Eberhard Klein, the chief investigator on the Mengele case, spent several hours with me reviewing the evidence his government had gathered over the years.

In New York, Eliot Welles of the B'nai B'rith Anti-Defamation

League opened up his organization's voluminous archives and files for my perusal.

In Washington, Opal Ginn, Jack Anderson's personal assistant, generously gave me months and months of paid leave to research the book.

In Tel Aviv, Mengele-hunter Menachem Russek patiently went over the results of the hunt, and shared with me his deepest and most private misgivings about the worldwide investigation. My thanks, also, to Isser Harel, former chief of the Mossad, for offering his perspective.

And in Asunción, members of the beleaguered—but still extraordinarily vibrant—Jewish community defied the Stroessner regime at great personal risk to share with me all their knowledge, tips, and suspicions about Mengele.

Of course, *Children of the Flames* would never have come into existence without the "children"—the surviving twins of Auschwitz. Meeting the twins was a fateful experience that changed my life forever. They received me with open arms, opened their homes and their hearts and their souls to me, and did all they could to make this book come to fruition. No story has ever moved me as much as theirs. Perhaps none ever will.

Eva Mozes Kor, more than any other human being, inspired me with a passionate need to track the Mengele story and get to know his twin victims. From her home in Terre Haute, Indiana, she functioned as a kind of "Command Central" for scores of surviving twins around the world. This book would not have been possible without the access to the network of twins Eva provided me with. I am also thankful to Vera Grossman, who introduced me to all the twins residing in Israel, and was able to convince them to trust me and open up to me. It must have been painful for her not simply to sit through interview after interview, but to translate from the Hebrew every heartrending word of their life stories for my benefit.

Several friends helped me with the editing of this book. I am grateful to Washington attorney Elaine Deering, who was always at my side, making sure the writing was fluid, and providing innumerable suggestions for improving the text. Sy Fischbein and Jane Weinbrenner also assisted me greatly.

In New York, Peter Bloch, the editor of *Penthouse* and a personal friend, never lost faith in me or this project.

In Washington, Wayne Savage donated office space and computers, to enable me to write the book.

I would also like to convey my gratitude and deep affection for John Cotter and Richard Gooding, my former editors at the *New York Post*, who were so helpful and encouraging during the final—and most crucial—stages of this book.

I owe a great deal to my editor, Susan Leon, formerly with Atheneum, and now with William Morrow. Susan approached this project with passion and saw it through, thanks to an inordinate amount of patience. She brought to it enthusiasm, but always tempered by a scholarly insistence on the importance of factual detail. It was Susan who saw the poetry amid the horror, and sensed the lyricism in the twins' voices even as they expounded on the most trivial everyday events—a good meal, a wedding, a blood test, an excursion to the beach. She provided strong editorial guidance thanks to her unflinching commitment to this book and her vision of what it could be, as well as her own quasi-mystical bond with the *Children of the Flames*.

And finally, I thank Doug Feiden, for being my friend and protector throughout the production of *Children of the Flames*.

–L.M.L.

The research and writing of what finally became *Children of the Flames* has almost as long a history as the infamous Auschwitz death camp itself. My husband Alex Shlomo Dekel's hunt for Dr. Josef Mengele was a self-imposed, lifelong mission. He wanted not only to locate the Doctor and bring him to trial, but to write a book encompassing the scope of the present work—Mengele's inhuman medical experiments on the twins and others. We began at least four drafts over a period of ten years in which we hoped to arouse the world's conscience.

When he died in 1983, I was left heir to his research. I turned to his dear friend, attorney Leon Charney, with my shopping bag of Alex's evidence and detailed files, including Mengele's fingerprints and SS records. Despite my grief, I was determined that Alex's research would not be left unpublished. Leon agreed with me and immediately contacted Pam Bernstein of the William Morris Agency, who was most sympathetic in her understanding that all that precious material could not go to waste. In September 1984, we both read Lucette Lagnado's article on Mengele's twins in *Parade* and decided to ask her if she would be interested in collaborating on a book that was aimed at exposing the deadly Doctor, who was then living in South America. Lucette had also been considering writing a book of her own. As she and I had mutual interests in the subject (inasmuch as Alex's article in *Life* magazine in June 1981 had first uncovered many sets of twins residing in the United States, Canada, and Israel) we decided to expand the subject matter from the viewpoint of one victim to all of those who had been subjected to Mengele's experiments. My heartfelt thanks go to both Leon Charney and Pam Bernstein for their encouragement in bringing to light Alex's work of twenty years.

But many events changed the original urgency of the book, most significantly, the discovery of the Doctor's remains in Brazil. Since the focus, if not the scope, had shifted, the book had to be reconceived, and I am indebted to our editor, Susan Leon, for her help and vision. Lagnado's interest in and research on the twins expanded this new concept. I wish to extend my gratitude to her for her steadfastness and commitment to this project over the years.

I would like to express my thanks to Simon Wiesenthal, a close friend and colleague of my husband's, for his help in aiding Alex to gain entry into the Polish National Museum, where their files were opened for the first time to him.

I would also like to express my thanks to Rabbi Morton Rosenthal, Director, Latin American Affairs, B'nai B'rith Anti-Defamation League, who gave us unlimited access to the league's files on South America and neo-Nazi activities. Also, I wish to thank Eliot Welles of that same organization for his dedication and help in this regard.

Early on, Ambassador Meir Rosenne, Alex's colleague for many years in the service of the Israeli government, directed me to various people in Israel and to the Yad Vashem Memorial, where they gave unstintingly of their time.

Thanks, too, to William C. Mulligan, who began this project with me, and to Joseph Lovett of *20/20* who supplied me with a tape of Alex prepared shortly before his death and which provides the source for many of his quotes used herein.

Most of all, I wish to express my thanks to David Friend, Senior Editor of *Life* magazine, to whom Alex first brought the devastating story of the twins, for his constant support and encouragement on this project. His devotion will never be forgotten.

–S.C.D.

Notes

Authors' Note: Research for this book took place between January 1984 and September 1990. Many of the individuals cited below (some of whose names appear in the Dramatis Personae and in the Acknowledgments) were interviewed numerous times, both in their homes and offices in the United States, Europe, Israel, and South America, as well as via overseas calls. Because of the complexity of the material, some sources were contacted again and again to ascertain the accuracy of certain relevant facts and intriguing recollections, therefore making it impossible to date specific conversations listed below.

PAGE

1 MENGELE AND HIS CHILDREN

29 The townspeople recall . . . get ready to unload: Mrs. Julia Kane, former Günzburg native, now residing in White Plains, NY, personal interview, May 1986. Also, Hermann Abmayr, West German journalist who grew up in Günzburg, personal series of interviews, spring 1986.

29 Karl and Walburga Mengele . . . farm-equipment factory: Mrs. Julia Kane, personal interview.

29 They would sigh . . . "Mengele is coming": Ibid.

30 A quarter of a century . . . a different railway stop: Robert Jay Lifton, *The*

Death Doctors (London: Macmillan, 1986), p. 343. Nearly all survivors who encountered Mengele recall his look of delight while serving on selection duty. Like a little boy, he wore "a cheerful expression on his face . . . almost like he had fun . . . he was very playful," Lifton cites one survivor as saying.

31 As a youth . . . not especially studious: Josef Baumeister, a teacher in Günzburg who has delved into Mengele's past by studying his school records, personal interview, May 1985. "He was very average," declared Baumeister—an impression confirmed by a perusal of Mengele's mediocre school grades.

31 There was an innocence . . . savage deeds at Auschwitz: Interview with Dr. Hermann Lieb, Mengele family dentist and personal friend of Josef, personal interview, May 1985.

31 He looked less like a Nazi . . . SS captain: Fania Fénelon, *The Musicians of Auschwitz* (London: Michael Joseph), p. 159. Fénelon compares Mengele to the late French heartthrob Charles Boyer.

32 In the years before World War I . . . house with another family: Dr. Zdenek Zofka, writer and unofficial Günzburg historian, who grew up in the town, personal series of interviews, 1985–89.

32 At the end of the war . . . among the largest in Günzburg: Ibid.

34 Neighbors recall how Josef . . . outings with their parents: Hermann Abmayr, personal interview.

34 A few older residents . . . small pond near their house: Ibid. Abmayr conducted interviews of several older Günzburg residents on behalf of author, requesting any memories they might have of Mengele as a child.

34 His early school records . . . exceptionally good behavior: Mengele's school records, courtesy of U.S. Justice Department, Office of Special Investigations. Translated by David Marwell, historian, OSI.

34 Though a mediocre student . . . and high school years: Ibid.

35 One Günzburg woman . . . this impassioned craving: Mrs. Julia Kane, personal interview.

36 Dr. Zdenek Zofka, the unofficial historian . . . Wally came to visit: Zdenek Zofka, personal interview.

36 They nicknamed her "the Matador" . . . out of her way: Ibid.

36 In an unpublished autobiography . . . a new automobile: Ibid. Zofka was invited to review autobiographical memoirs of Mengele in possession of Bunte magazine in June 1985. He shared contents with author during series of interviews.

36 "I will always stay with you," he told her: Ibid.

37 When he was fifteen . . . always fatal in those days: Berlin Document Center file on Mengele.

37 He also developed . . . severe systemic infection: Ibid.

38 He failed several subjects . . . "and more ambitious": Mengele school records, courtesy OSI and David Marwell.

38 Josef and his friends dressed impeccably . . . and white gloves: Josef Baumeister, personal interview.

38 They didn't actually wear the gloves, but only held them, nonchalantly, in one hand: Ibid. This exquisite detail is confirmed by a glance at photographs of Josef from the period, including his wedding picture. There, too, he is shown only holding the white gloves.

38 It is ironic . . . solitary, self-effacing youth: Mengele autobiography, courtesy of OSI and David Marwell.
38 He depicted himself as monklike and ascetic . . . young men his age: Ibid.
38 Josef was certainly conscious . . . wished he were taller: Inge Byhan, *Bunte* magazine 1985 series on Josef Mengele, Part II, Bunte Archives 27/85, Munich, Germany. Handwritten translation by Elizabeth Bergmann.
39 Young Beppo's elite group . . . in the late 1920s: Josef Baumeister, Hermann Lieb, personal interviews.
39 Mengele's crowd was completely uninvolved . . . cars than politics: Ibid.
39 But Mengele and many . . . Greater Germany Youth Movement: Dr. Zdenek Zofka, personal interview.
39 The young people's branch of the Stalhelm . . . values of the earth and "Mother Germany": Zofka, personal interview.
40 In some ways . . . kibbutzim of Palestine: Ibid.
40 According to Zofka . . . gravitated toward this movement: Ibid.
40 He officially joined the *Stalhelm* . . . brothers followed suit: Berlin Document Center files on Josef, Karl, and Alois Mengele.
40 His closest childhood friend Hermann Lieb . . . political movements of the day: Hermann Lieb, personal interview.
40 With chilling prescience . . . "in the encyclopedia" Gerlad Posner and John Ware, *Mengele: The Complete Story* (New York: McGraw-Hill, 1986), p. 5.
40 He had been . . . the friendliest and the most humane: Josef Baumeister, personal interview.
41 The centerpiece of Hitler's speeches . . . driving force of Mengele's existence: Adolf Hitler, *Mein Kampf*, trans. Ralph Manheim (New York: Houghton Mifflin, 1943). Many pages of *Mein Kampf* consist of Hitler's ramblings on the importance of German racial superiority. See Vol. I, Chapter XI, Nation and race, p. 285, and Vol. II, Chapter II, Folkish State and Racial Hygiene, p. 403.
41 Hitler called for the elimination of . . . especially the Jews: Entire libraries have been filled with books analyzing Hitler's plans for the annihilation of the Jews, the Slavs, the Gypsies, and all other so-called "inferior" races. The reader is urged to consult his blueprint for the killing of the Jews as described in *Mein Kampf*, which first appeared in 1925. See pp. 300–325; 401–3; 622–40.
41 Even as Hitler and his supporters . . . great shows of profundity: Max Weinreich, *Hitler's Professors* (New York: Yiddish Scientific Institute—YIVO, 1946), p. 9.
41 Mengele's chosen fields of anthropology and genetics were especially influenced by the racist theories of Nazi dogma: Ibid., pp. 31–35.
41 His university records suggest . . . compared to his peers. Mengele's University of Munich records, 1930–35, courtesy of University of Munich, Munich, West Germany.
42 Friends from the period . . . diligent Mengele was: Dr. Kurt Lambertz, interview by telephone, August 1986.
42 As his autobiographical writings indicate . . . obtaining a medical degree: Mengele autobiography, courtesy of Zdenek Zofka. Several of the scholars who read through Mengele's often excruciatingly tedious notebooks were struck by one

anecdote that suggests that even early on, he approached his studies with a kind of demonic fervor. Like other medical students, Mengele was required to perform dissections. But there were never enough human corpses to go around. Mengele, who was also studying zoology, recalls pestering officials at the Munich zoo for permission to dissect dead animals, especially apes and monkeys. One day, the call he had been awaiting finally came: The director of the zoo was on the line. An ape had died just a little while before. Would Mr. Mengele be interested in examining it? Mengele recalls how he ran out of his room, jumped into his car, and raced toward the Munich zoo, thoroughly excited at the notion of performing an autopsy on an animal "while it was still warm."

42 The social Darwinists . . . personal and social problems were inherited: Daniel J. Kevles, *In the Name of Eugenics* (New York: Alfred A. Knopf, 1985), pp. 20–21.

42 The social Darwinists espoused a program . . . only the "best" people survived: Ibid., p. 20.

42 The theories of the social Darwinists . . . Germany and the United States were fostering a eugenics movement: Ibid., pp. 73–75 on American eugenics movement. See also pp. 116–18. "In the early years of the Nazi regime, most mainline eugenicists in the United States and Britain could not know—and likely did not want to know—that a river of blood would eventually run from the sterilization law of 1933 to Auschwitz and Buchenwald," Kevles observed starkly.

42 In Washington, for instance . . . "contaminate American blood": Ibid., p. 97.

42 While arguing for passage . . . "inferior to Nordics": Ibid., p. 97. Kevles quotes New York congressman Samuel Dickstein, a Jewish member of the House Committee on Immigration and Naturalization, as saying, "If you had been a member of that committee, you could not help but understand that they did not want anybody else in this country except the Nordics."

43 The Germans, however, carried racial science farther . . . its extreme anti-Semitic component: Ibid., pp. 117–18.

43 His writings suggest . . . "the fate of mankind": Dr. Zdenek Zofka, personal interview.

43 His account of that period . . . too busy even to marry: Ibid.

43 But none did so with more fervor . . . acclaimed racial scientist of his day: Dr. Klaus Dieter Thomann, "Racial Hygiene and Anthropology: Professor Verschuer's Twin Careers," *Frankfurter Rundschau*, May 21, 1985, p. 8.

43 Since the early 1930s, Verschuer . . . which specialized in racial studies: Ibid.

44 He promptly applied for and was given . . . the rising star of Germany's eugenics movement: Ibid.

44 Frankfurt in the 1930s . . . new Nazi regime: Dr. Klaus-Dieter Thomann, personal interview, June 1985, Frankfurt, West Germany.

44 The city's ministry of health . : . carried any other "genetic disease": Thomann, *Frankfurter Rundschau*, p. 8.

44 Mengele joined his institute colleagues . . . charged was indeed a Jew: Benno Müller-Hill, *Todliche Wissenschaft: Die Aussonderung von Juden, Zigeunern und Geisteskranken 1933–1935* (Hamburg: Rowolt, 1984), p. 39. Benno Müller-Hill, himself a scientist, was able to interview the son of Verschuer

about his father. Whether Verschuer's child was more candid with Benno Müller-Hill is debatable, however. Asked about his father's infamous racial evaluations of individuals suspected to be Jewish, he merely replied, "My father also helped Jews with his evaluations—after the War, he mentioned receiving thank-you notes from America."

44 Within a few years, Mengele's judgments . . . deportation to the death camps: Thomann, *Frankfurter Rundschau*, p. 8.

45 His job as an assistant to Verschuer . . . Nazi racial medicine: Ottmar, Katz, West German journalist and Mengele specialist, personal interview, Munich, May 1985.

46 The professor provided the critical link . . . grotesque experiments on children: Ibid. Katz, a magazine journalist who has spent years studying Mengele, and has a roomful of thick files on him, is passionate about this point: that Otmar von Verschuer was the man who led to Mengele's downfall. "I will not defend Mengele—he was a murderer. But he was only the disciple. The master was Verschuer," says Katz.

46 In the 1920s, when Verschuer . . . almost exclusively on twins: Thomann, *Frankfurter Rundschau*, p. 8.

46 Verschuer did, however, instill . . . "in vivo" research, could be: Thomann, personal interview, May 1985.

47 As a result of his diligence, Mengele . . . often stayed for dinner: Benno Müller-Hill, *Todliche Wissenschaft*. Also, Verschuer's daughters discussed Mengele's visits to their father's house with Vera Grossman Kriegel, one of Mengele's twin guinea pigs. Vera Grossman Kriegel, personal interview, May 10, 1987.

48 Verschuer, who was in constant communication . . . frequent tribute to Hitler in his various publications: Thomann, *Frankfurter Rundschau*, p. 8.

48 His articles routinely praised the Führer . . . from German society: Ibid, p. 8.

48 Ultimately, Verschuer helped give the Final Solution . . . intellectual respectability: Dr. Klaus-Dieter Thomann, *Medizin im Faschismus* (Medicine During the Fascist Era) (Berlin: Veb Verlag Volk Und Gesundheit, 1985), p. 65.

48 Although he never held high office in the Reich . . . carried considerable weight: Ibid., p. 65.

48 As a perceptive American investigator . . . "the favor of the Nazi tyrants": Memorandum by Manfred Wolfson, chief research analyst, Office of Chief of Counsel for War Crimes, Berlin Branch, November 7, 1946.

49 That same year . . . while on holiday: Kurt Lambertz, telephone interview.

49 Mengele was assigned . . . skiing and hiking: Berlin Document Center file.

49 According to Dr. Kurt Lambertz . . . of the unit's commanding officer: Lambertz interview.

49 The personality clash . . . between the two: Ibid.

49 His first assignment . . . applications for German citizenship: Berlin Document Center file.

49 In Hitler's view of the world . . . wherever they might be living: David Marwell, OSI, personal interview.

49 In June 1941 . . . the Waffen SS: Berlin Document Center file.

50 The task of choosing . . . told friends and colleagues: Ottmar Katz, personal interview.

50 He ended up being awarded the Iron Cross, First Class . . . under enemy fire, Posner and Ware, *The Complete Story*, p. 17.

50 As one commendation . . . "face of the enemy": Lifton, *The Nazi Doctors*, p. 340.

50 The SS reassigned him . . . overseeing the concentration camps: Berlin Document Center file.

51 Since 1939, medical-research projects . . . on human subjects: *Encyclopedia of the Holocaust*, (New York, MacMillan, 1990) Vol. 3, p. 957.

51 To find a cure for typhoid fever . . . with various serums: Ibid., p. 953.

51 At Dachau and other camps . . . these deadly maladies: Ibid. pp. 957–965.

51 Nineteen forty-two was also . . . method of mass sterilization: Ibid., p. 964.

51 It was at Auschwitz that a certain Dr. Horst Schumann . . . the desired sterilization: Ibid., p. 965. In one of the most gruesome chapters of the Holocaust, Schumann is believed to have exposed hundreds of people to massive doses of radiation aimed at their reproductive organs. "Most of them were sent to the gas chambers afterward, since the radiation burns they suffered made them unfit for work."

52 Verschuer even helped Mengele win grants to undertake two research projects at the camp: Lifton, *The Nazi Doctors*, p. 341. See also Benno Müller-Hill, *Wissenschaft*, pp. 23 and 72. Although it is unclear just how Mengele obtained his position as an SS doctor at Auschwitz, the preponderance of evidence suggests Verschuer was closely involved in the transfer, and may have actually facilitated it, using his boundless connections. Clearly, Verschuer had every intention of taking advantage of Mengele's work at Auschwitz. And we know from survivors such as Dr. Miklos Nyiszli, a pathologist who worked closely with Mengele at Auschwitz, that Verschuer succeeded in doing so, perhaps beyond his wildest dreams. According to Nyiszli, the diligent young Mengele regularly dispatched blood samples and body parts to his mentor in Berlin. The ever-resourceful Verschuer managed to turn Mengele's laboratory at Auschwitz into a satellite to Kaiser-Wilhelm.

2 Auschwitz Movie

54 What is universally known today as Auschwitz is in fact something of a misnomer: Much of this book takes place not at Auschwitz but in the adjoining camp of Birkenau. Auschwitz was actually a series of camps spread out over forty kilometers. There were thirty small labor camps, where slave-inmates toiled away at everything from fish breeding and vegetable gardens to poultry farms and munitions factories. The most infamous of all the camps was Birkenau, or Brzezinka, where the majority of the killing machines were actually situated. See *Encyclopedia of the Holocaust*, Vol. I, pp. 107–121.

54 Although the slaves . . . many of them were executed: Ibid.

54 It was Birkenau where the ovens . . . so many of them inevitably perished: In the course of interviewing nearly fifty twins who had been subjected to experiments by Mengele, I found that nearly all of them objected to my using the name "Auschwitz" interchangeably with "Birkenau." Rather, they were pas-

sionate, almost pedantic, in their insistence that I be precise in distinguishing between the two camps.

55 Every morning, at the crack of dawn . . . scanning the new arrivals: Lifton, *The Nazi Doctors*, p. 342.

55 Standing there in his perfectly tailored SS uniform . . . guests arriving at his home: "Report on the Auschwitz Concentration Camp," unsigned affidavit, Office of Chief of Counsel for War Crimes, August 20, 1946. This document consists of a twenty-one page first-person account by a nurse who worked with Mengele, and dubbed him "one of the most refined butchers of the mass executions at the Birkenau death camp." Mengele "succeeded in annihilating men, women and children with the manners of the most perfect gentleman," wrote the unnamed nurse.

55 Often, he whistled . . . his favorite Puccini opera: Ibid., p. 16.

55 Mengele even engaged some . . . how they were feeling: Ibid., see pp. 9 and 15. "Several times, we noticed the hypocritical manner in which the grim doctor Mengele treated women and children alighting from the train.

" 'Madam, take care, your child will catch cold.'

" 'Madam, you are ill and tired after a long journey, give your child to this lady and you will find it later in the children's nursery.' "

55 Occasionally, Mengele pulled aside inmates . . . relatives back home: Testimony of Isaac Egon Ochshorn, United Nations War Crimes Commission, Statement by Ochshorn on Massacres of Jews in Concentration Camps, U.S. National Archives, September 1945, p. 3.

55 He seemed to take a special pleasure . . . and urging everyone to visit: Ibid., p. 3.

55 SS guards were ordered . . . bring them immediately to Mengele: Sara Nomberg-Prytyk, *Auschwitz: True Tales from a Grotesque Land* (Chapel Hill: University of North Carolina Press, 1985) pp. 89–90. Mengele "loved to single out those who had not been created in 'God's image' " wrote Nomberg-Prytyk, who worked as a nurse in Mengele's sham infirmary. She recalled how he loved to bring in human "freaks" he had collected from the selection lines—a woman with two noses, one with donkey's ears, and a little girl "who had the wool of a sheep on her head instead of hair." All became members of Mengele's menagerie.

55 But most important . . . were the twins: Magda and Zvi Spiegel, twins at Auschwitz, personal interviews, March 1984.

56 And if the twins were just infants . . . and look after them: Rosie Rosenbaum, mother of twins Judith and Ruthie Rosenbaum, personal interview, March 1984. Also, Rosalea Csengery, mother of twins Lea and Yehudit Csengery, personal interview, March 1984. Both Mrs. Rosenbaum and Mrs. Csengery were permitted to accompany their twins, because the children were so small. Both women said, however, that this was the rare exception rather than the rule. In fact, because there were so many motherless infants, these beleaguered women ministered to the other children as well.

57 But unlike the other prisoners . . . both their clothes and their long hair: Zvi Spiegel, twin at Auschwitz, personal interviews, March 1984 and May 1988.

57 These differences . . . "Mengele's Children": Zvi Spiegel, Ibid.
58 But the new twins also learned . . . kingdom of death were guaranteed: Eva Mozes, twin at Auschwitz, personal interview, March 1984. See also Gisella Perl, *I Was a Doctor in Auschwitz* (New York: International Universities Press, 1948), pp. 132–33.
58 Mengele made sure that . . . by Auschwitz standards: *Doctor in Auschwitz*, p. 132. Perl seems to have been almost jealous of the twins' status at Auschwitz under Mengele. "To be a twin in Auschwitz seemed the maximum of good fortune," Perl wrote. "They were the chosen ones, the highest caste, the spoiled darlings of the SS doctor."
58 Because they "belonged" to Mengele . . . would dare lay a hand on them: Zvi Spiegel, personal interview.
58 In addition to keeping their clothes and hair . . . food rations than the other prisoners: Eva Mozes, Peter Somogye, twins at Auschwitz, personal interviews 1985–88.
58 Although all the twins say . . . enabled them to survive: Mozes, Spiegel, Zvi Klein, Menashe Lorinczi, and several other twins mentioned their access to some food during their interviews.
59 Most important, the twins . . . random selections that adult prisoners faced: Perl, *Doctor in Auschwitz*, pp. 132–33.
59 A superior's evaluation . . . "maturity and strength." *Wolfson Report*, citing SS personnel assessment of Mengele by SS Hauptsturmführer Mattes, dated August 19, 1944.
59 And one doctor who served with Mengele . . . the death camp: Posner and Ware, *The Complete Story*, p. 42.
59 This perception of Mengele as more hardworking . . . their memoirs and testimonies: Dr. Miklos Nyiszli, *Auschwitz: A Doctor's Eyewitness Account* (New York: Frederick Fell, 1960), p. 81.
59 Mengele's experimental barracks . . . the Nazi hierarchy: Perl, *Doctor in Auschwitz*, p. 132.
59 Verschuer, from his position . . . and the two men corresponded regularly: *Wolfson Report*, p. 3. Also see Nyiszli, *Auschwitz*, pp. 82, 84–86.
60 The detailed questionnaires . . . Mengele's work: Zvi Spiegel, personal interview.
61 In addition, it is believed that Mengele . . . the depleted German Army: Nyiszli, p. 80.
61 The ultimate goal . . . and rule the world: Ibid., p. 80. Also see Lifton, *Nazi Doctors*, p. 359.
61 He installed a small funished office . . . child guinea pigs: Magda Spiegel, personal interview, March 1984.
61 He decreed that guards . . . look after the children: Zvi Spiegel, personal interview, May 1987.
61 If a twin died . . . demanding an explanation: Eva Mozes, personal interview, March 1986.
61 He performed selections . . . hundreds to be killed: Perl, *Doctor in Auschwitz*, pp. 129–30.
61 He oversaw . . . alas, little care was provided: Nomberg-Przytyk, p. 95. Perl,

Doctor in Auschwitz, pp. 70–71. In her inimitable style, Perl calls the Auschwitz infirmary a "ghastly Nazi joke . . . the screams of the patients seemed to give Dr. Mengerle [sic] a perverse pleasure."

61 Typically, Mengele gave . . . "prepared" for experiments: Zvi Spiegel, interview, May 1987.

62 Special trucks emblazoned . . . to Mengele's laboratories: Unsigned report, Office of Chief Counsel for War Crimes, August 10, 1946, p. 9. The unidentified witness recalls:

"Ah, if the Red Cross lorry could but speak, what scenes and what unheard of conversations it could relate! It was used for the transporting of invalids, of children, of old people, of pregnant women who supposedly were taken to the hospital, but who 9 times out of 10 were thrown directly into the trenches or then shot . . ."

62 Depending on . . . or at Auschwitz proper: Eva Mozes, series of personal interviews, 1984–89.

62 In one laboratory . . . and painful—experiments: Ibid.

62 One Mengele lab the twins never saw . . . of a crematorium: Nyiszli, *Auschwitz*, pp. 59–61.

62 There, an assistant . . . course of experiments: Nyiszli, *Auschwitz*, pp. 123; 155–57; 159–62; 82–85.

62 These tests may have been . . . express request: Müller-Hill, *Todliche Wissenschaft*, p. 72.

62 The youngest children . . . frightening procedure: Solomon Malik, twin at Auschwitz, March 1984.

62 Although Mengele was . . . by his assistants: Zvi and Magda Spiegel, personal interviews, March 1984.

62 Most had profound misgivings about their work: Magda Spiegel, personal interview, March 1984.

62 In spite of . . . helped save some lives: Ibid.

63 He looked the twins over . . . other unusual conditions: Zvi Spiegel, personal interview, March 1984.

63 Then, he would demonstrate . . . or draw blood: Moshe Offer, twin at Auschwitz, personal interview, March 1984.

63 As if he were their old family physician . . . the youngsters: Ibid.

64 He intuitively understood . . . only if they were cooperative: But if they weren't cooperative, he was thoroughly menacing. The best witnesses to Mengele's behavior toward the children were the handful of mothers he allowed to be present during the experiments. Mengele expected the mothers to keep their children quiet and make them cooperative. The mothers knew that if their twins cried or misbehaved, Mengele would inflict severe punishments. Mrs. Rosalea Csenghery, personal interview, March 1984.

65 Mengele's experiments . . . followed few scientific principles: Perl, *Doctor in Auschwitz*, p. 133. Perl calls Mengele's medical work a "sham," and denounces him for using the children as "guinea pigs." See also Nyiszli, *Auschwitz*, pp. 78–79. Lifton, however, cites many survivors who assert that Mengele saw his work as genuine scientific research. Lifton, *Nazi Doctors*, pp. 368–69.

65 The eye studies were especially gruesome: Hedvah and Leah Stern, personal interviews, March 1984.
65 A major focus . . . twins' hair and eyes: Zvi Spiegel, personal interview, May 1988.
65 His fascination with hair . . . wear it long: Ibid.
65 Their hair was continually . . . of their sibling: Ibid.
65 The eye studies . . . by Mengele at Auschwitz: Wolfson report, pp. 3, 6.
65 At times, Mengele's assistants . . . they used needles: Hedvah and Leah Stern, personal interviews.
66 He loved to sit and chat with them: Vera Blau, twin at Auschwitz, personal interview, March 1984. Also Peter Somogyi, twin at Auschwitz, personal interview, June 1987.
66 Mengele especially seemed to dote . . . completely dependent on him: Vera Blau, personal interview.
67 There was a little boy . . . to Mengele himself: Ibid.
67 He was a dark child . . . a gentle disposition: Ibid.
67 Mengele came to the barracks . . . and chocolates: Ibid.
67 When asked his name . . . "is Mengele": Ibid.
67 Zvi Spiegel . . . entertain the youngsters: Zvi Spiegel, personal interview.
67 An avid sportsman . . . composed of the twins: Zvi Klein, twin at Auschwitz, personal interviews, March 1984 and May 1987.
69 But the older twins . . . of his perverse experiments: Mrs. Rosie Rosenbaum, mother of twins, interview, March 1984, and Magda Spiegel, twin at Auschwitz, interview, March 1984.
69 However, in performing . . . high-minded scientific goals: Lifton, *Nazi Doctors*, p. 365.
69 If there was ever . . . Otmar von Verschuer: Ottmar Katz, personal interview, May 1987.
70 Bizarre psychological tests . . . were continually made: Vera Grossman, twin at Auschwitz, personal interviews, May 1987.
70 For instance . . . with or without his twin: Eva Mozes, personal interviews, 1984–89.
70 And finally, horrible, murderous operations *New York Times*, February 7, 1985, pp. 1, 12.
70 Mengele would plunder . . . organs and limbs: Ibid. Also, Moshe Offer, personal interviews, re: twin Tibi, March 1984.
70 He also attempted . . . of some twins: Eva Mozes, personal interview.
70 Female twins . . . males were castrated: *New York Times*, February 7, 1985, p. A12.
71 Several twins believe . . . twins mate: Zvi Spiegel, personal interview, March 1984. Also, Eva Mozes, personal interview.
71 The final step . . . have their organs analyzed: The consensus of the surviving twins of Auschwitz, all of whom came to the camp relatively late in the war, is that Mengele conducted his experiments in stages. They believe they came out alive simply because Mengele did not get to the "final" stage of his experiments—their murder—because of the war's fortuitous end. They cite the fact that hundreds and hundreds of twins who preceded them perished.

71 Occasionally, though, if . . . phenol to the heart: Nyiszli, *Auschwitz*, p. 83.
71 An autopsy would be performed . . . Verschuer's Berlin institute: Ibid., p. 84.

3: The Angel of Death

73 By the summer . . . selections at Auschwitz: Normberg-Przytyk, *Grotesque Land*, pp. 79, 81. "Day and night, trainloads of people were unloaded on the ramps. Most of them went directly to the gas-chambers," wrote Normberg-Przytyk. "That July and August, the weather was very hot and stuffy. . . . Every day of that macabre last summer of Hitler's reign, twenty-thousand people were killed in Auschwitz. The crematorium was unable to burn all of the dead . . ."
73 Hardly anyone now arriving . . . labor camps: Ibid.
73 The war was now . . . destroy their Jewish victims: Ibid.
73 Hitler's army . . . massive German defeats: David Marwell, OSI, personal interview.
73 The Hungarian Jews . . . in Eastern Europe: Lucy Dawidowicz, *The War Against the Jews* (New York: Bantam Books, 1976), pp. 513–17.
74 Although formally allied . . . for most of the war: Ibid., pp. 515–16. Hungarian leaders even "rejected German demands to introduce yellow badges for Jews and deport them to Poland."
74 But in 1944 . . . to the death camps: Ibid. p. 516.
74 By the summer of 1944 . . . deported to Auschwitz: Ibid., p. 517.
75 The preponderance of evidence . . . of his heinous deeds: It was impossible to track precisely when the term "Angel of Death" became Mengele's popular nickname. But just after the war, two survivors who were also doctors, Gisella Perl and Miklos Nyiszli, both published detailed accounts of life under Dr. Mengele—and neither referred to him using the term "Angel of Death." It is reasonable to suppose that if it had indeed been his nickname at the camp, one or both would have been sure to mention it. Olga Lengyel's *Five Chimneys*, another survivor's powerful first-person story of Auschwitz and Mengele, does refer to a "blonde angel"—but she means Irma Greze, Mengele's assistant. Certainly, however, by the 1970s and the 1980s, the nickname was completely fused in the public's mind with Mengele's dangerous, elusive persona.
76 The Angel of Death . . . the Old Testament: *Encyclopedia Judaica* (New York: Macmillan, 1971), Vol. 1, pp. 57–59.
76 By chilling coincidence . . . "of excellent repute": *The Encyclopedia of Judaism* (New York: Jewish Publishing House, 1989), p. 954.
76 The ancient spiritual leader . . . "doctors to assist him": *Encyclopedia Judaica*, Vol. 1, pp. 58–59.
76 Even if the Almighty . . . would doubtless have shared: *Encyclopedia of Judaism*, p. 954.
76 Like the spirit *Malach Hamavet* . . . a human life: Ibid., pp. 953–54.
77 Although the twins were . . . did not get out of hand: Perl, *Doctor in Auschwitz*, p. 94.
77 One especially hot . . . from the cattle car: Nomberg-Przytyk, *Grotesque Land*, p. 105. This amazing story is told by Ms. Nomberg-Przytyk in a chapter hauntingly titled "The Dance of the Rabbis." One finds in her book the clear sug-

gestion that Mengele was a rabid anti-Semite, who reveled in humiliating the Jews. According to Nomberg-Przytyk, Mengele even timed especially large selections to coincide with Jewish High Holidays—a recollection that is echoed by several twin survivors.

77 Mengele looked . . . to the gas chambers: Nomberg-Przytyk, *Grotesque Land*, pp. 105–6.

77 First, he ordered . . . and sing: Ibid., p. 106.

77 Then, Mengele commanded them to dance: Ibid., p. 106.

77 Mengele's zeal for his work clearly impressed his superiors, who showered him with accolades that terrible summer: Wolfson report, re: August 19, 1944, SS evaluation of Mengele, p. 2.

77 He was now at the height of his powers . . . several thousand female inmates: Lifton, *Nazi Doctors*, p. 349. Also see Olga Lengyel, *Five Chimneys* (New York: Howard Fertig, 1983), p. 144. Precisely when Mengele became chief doctor of Birkenau is unclear, however. What is irrefutable is that during the summer of 1944, Mengele was the most detested man at Birkenau because of his passion for selections.

77 Although there were Nazis who held higher rank . . . feared as Dr. Josef Mengele: Lengyel, *Five Chimneys*, p. 144.

78 In the terrible summer of 1944, Birkenau was also crammed . . . Hungarian women: Perl, *Doctor in Auschwitz*, p. 80.

78 Because of the stepped-up number of transports . . . Jews who kept arriving: Nomberg-Pzrytyk, *Grotesque Land*, p. 81.

78 Mengele was frequently accompanied . . . Irma Grese: Lengyel, *Five Chimneys*, p. 144.

78 The "Blond Angel" . . . couple: Lengyel, *Five Chimneys*, p. 147.

78 Only eighteen, Irma . . . of Jewish women: Ibid., pp. 147–48. (Note: Lengyel incorrectly identifies her age as twenty-two. But at the Belsen trial, after the war, Grese was revealed to have been just eighteen during her service at Auschwitz.)

78 She gloated over the inmates . . . at her mercy: Ibid.

78 She cracked her leather whip . . . female SS guard: Perl, *Doctor in Auschwitz*, p. 62.

78 To Mengele, of course . . . utterly deferential: Belsen trial records, U.S. National Archives.

78 Much as they loathed . . . found him attractive: Fénelon, *Musicians of Auschwitz*, p. 159.

78 As he inspected them . . . and attempting a smile: Ibid. Fénelon writes, "Goodness, he was handsome. So handsome that the girls instinctively rediscovered the motions of another world, running dampened fingers through their lashes to make them shine, biting their lips, swelling their mouths, pulling at their skirts and tops. Under the gaze of this man one felt oneself become a woman again."

78 But despite their physical frailty . . . Mengele's sexual magnetism: Ibid.

79 Much as he was able . . . inmates feel at ease: Nomberg-Przytyk, *Grotesque Land*, p. 56.

79 Several of the women . . . chronic ailment: Ibid.

79 What these women . . . hastening their own deaths: Ibid.

79 Shortly after arriving . . . heads were shaved: Perl, *Doctor in Auschwitz*, pp. 42–44.
79 The women were then given . . . rags to wear: Ibid. p. 45.
79 Their shoes were . . . and one high heel: Ibid., p. 46.
79 The effect . . . their Aryan guards: Ibid., pp. 42, 47. "Sisters, friends did not recognize each other any longer," Perl writes hauntingly, "and the prettiest girls and most beautiful women looked like a bunch of grisly monsters, ridiculous, and sub-human."
79 Mengele encountered such a woman in Ibi Hillman: Perl, *Doctor in Auschwitz*, p. 138.
79 Tall, blond, and statuesque . . . village in Transylvania: Ibid., p. 137.
80 When Ibi removed . . . transfixed: Ibid., p. 138.
80 In a loud voice . . . sinister gynecological experiments: Ibid., p. 138.
80 Few women survived Block Ten: *Ochshorn Statement*, p. 16.
80 The beautiful young girl looked like a shriveled old woman: Perl, *Doctor in Auschwitz*, p. 138.
80 This is not . . . sent to die: Lengyel, *Five Chimneys*, p. 102.
80 Mengele even boasted . . . having these women killed: Nomberg-Przytyk, *Grotesque Land*, p. 69.
80 But Mengele fluctuated erratically in his policy: Perl, *Doctor in Auschwitz*, p. 84.
80 Some weeks, he issued . . . every consideration: Ibid.
80 Other weeks, he ordered them killed immediately: Ibid.
80 When he first met . . . her at length about her condition: Lengyel, *Five Chimneys*, 102–3.
80 He asked dozens . . . a concerned physician: Ibid.
80 In posing the questions . . . who overheard such exchanges: Magda Spiegel, personal interview, March 1984.
81 Mengele's Jewish assistants . . . discovered: Perl, *Doctor in Auschwitz*, p. 71.
81 In an effort . . . performed crude abortions: Ibid. p. 81.
81 When a pregnancy was too advanced . . . been born dead: Ibid., pp. 81–82. "No one will ever know what it meant to me to destroy these babies," Perl wrote sadly. "I loved those newborn babies not as a doctor but as a mother and it was again and again my own child whom I killed to save the life of a woman."
81 In *I Was a Doctor in Auschwitz* . . . the dwarfs and midgets: Perl's memoirs of life at Auschwitz are crammed with anecdotes about the abominable Dr. Mengele. The reader is directed to pp. 124–25; 120–22; 110–11.
81 Mengele was "so proud" . . . Perl wrote with emotion: Ibid., p. 133.
82 Several thousand Gypsies . . . by the crematoriums: Menashe Lorinczi, twin at Auschwitz, personal interview, March 1984.
82 At any time . . . to their deaths: Ibid.
82 Their encampment looked like a vast playground: Christian Bernardac, *L'Holocauste Oublié* (Paris: Editions France Empire, 1979), p. 174. Before slaughtering them, the Nazis lavished them with a strange gift: They built the Gypsy children a wooden merry-go-round.
82 The adults exchanged stories, sang songs, even danced: Ibid., p. 161.
82 Some of the old melodies . . . would stop to listen: Ibid.

82 He appeared fond . . . in the compound next door: Lifton, *Nazi Doctors*, p. 375.
82 "Uncle Mengele! Uncle Mengele!" they cried out.: Ibid.
82 Tests were being conducted . . . watchful for more: Nyiszli, p. 82. See also Lifton, *Nazi Doctors*, p. 348; Lifton notes that Mengele even had a separate office right in the Gypsy camp "for his work with twins."
82 There was one little boy . . . the Gypsy camp: Nomberg-Pzrytyk, *Grotesque Land*, pp. 82–83.
82 Mengele had him dressed . . . looked strangely regal: Ibid., p. 83. "The little boy was a beauty," Nomberg-Pzrytyk wrote. "He was dressed in a gorgeous white uniform, consisting of long pants with an ironed-in crease, a jacket adorned with gold buttons, a man's shirt, and a tie. We stared, as if bewitched at that beautiful child."
82 Sometimes he would ask . . . or sing a melody: Ibid.
83 Mengele knew that conditions . . . the death rate was mounting: Some scholars believe that Mengele actually tried to intercede to save the Gypsies. See Lifton, *Nazi Doctors*, p. 379, where he writes that "Mengele . . . objected to specific policies—the destruction of the Polish intelligentsia and the annihilation of the Gypsy camp."
83 Little girls danced to the music: *L'Holocauste Oublié*, p. 161.
83 The order finally came down from Berlin on August 2, 1944: *Encyclopedia of the Holocaust*, Vol. II, p. 636.
83 At seven o'clock . . . camp was sealed shut: Ibid.
83 Toward the end . . . his little mascot: Normberg-Przytyk, *Grotesque Land*, p. 84.
83 Hand in hand, they . . . and showed him the way inside: Ibid. The eerie tale of Mengele and the little Gypsy boy has been told by several survivors. Accounts vary of what happened those last minutes with the child. Ms. Normberg-Przytyk, however, claims that Mengele "pushed him into the gas-chambers with his own hands."
83 Efficient as ever, Mengele . . . were being conducted: Nyiszli, *Auschwitz*, p. 156.
84 They were gassed . . . would not be cremated: Ibid.
84 The corpses were duly delivered . . . as ordered: Ibid., pp. 159–62.
84 That same night . . . discussing the findings: Ibid., p. 162.
84 A few weeks later . . . wife, Irene, arrived: Posner and Ware, *Complete Story*, pp. 54–55.
84 This time she left the child behind in Germany: Ibid., p. 54.
84 The couple went swimming . . . where they picked berries: Ibid., p. 54.
84 Irene's Auschwitz holiday . . . to be hospitalized: Ibid., p. 56.
84 In an entry she made . . . a bit depressed: Ibid., p. 55.
84 One of Mengele's greatest fears . . . his research material: Dr. Ella Lingens, prisoner doctor at Auschwitz, personal interview, Vienna, May 1985.
84 One day, he invited . . . reports he had compiled: Ibid.
84 As Lingens leafed through . . . hands of the Bolsheviks?: Ibid.
84 And indeed that fall . . . to be rapidly approaching: Hedvah and Leah Stern, Menashe Lorinczi, twins at Auschwitz, personal interviews, March 1984.
86 Later that fall . . . deserted Gypsy barracks: Nearly all of the twins, even the

younger ones, remember the move, and how frightened they were by it. Menashe Lorinczi, Judith Yagudah, personal interviews.

86 The children were afraid . . . again for extermination: Ibid., Lorinczi.

4 The Angel Vanishes

88 For the inmates . . . hope and celebration: Normberg-Przytyk, *Grotesque Land*, p. 123.

88 At previous New Year's . . . for the Nazis: Ibid.

88 A furtive party . . . who were also prisoners: Ibid., p. 123.

89 They scurried around destroying . . . of their crimes: Ibid., p. 127.

89 Many of the crematoriums . . . hierarchy in Berlin: *Holocaust Encyclopedia*, Vol. 1, p. 116. See also Nyiszli, *Auschwitz*, p. 228.

89 Between December and January, for instance, 514,843 articles of men's women's and children's clothing were shipped out, according to an official Nazi report: Auschwitz Museum Archives, Auschwitz, Poland.

89 Entire storerooms remained . . . of the millions: *Holocaust Encyclopedia*, p. 117.

89 One morning . . . moved to another crematorium: Nyiszli, *Auschwitz*, p. 228.

89 The German Army . . . in preparation for more "work": Ibid.

89 Mengele apparently took steps to safeguard his work: Posner and Ware, *Complete Story*, p. 89. The evidence is not clear-cut, but in interviews both with authors Posner and Ware and to *Bunte*, Mengele's son, Rolf, has suggested his father did ferret away some of his scientific work with family members, who gave it back to him when he left Germany for good.

89 One prisoner did notice . . . a waiting car: Ibid., p. 58.

89 He left the camp so stealthily . . . Mengele was gone: Nyiszli, *Auschwitz*, p. 235.

90 Eager to avoid . . . to destinations unknown: Nomberg-Przytyk, *Grotesque Land*, p. 128.

90 With defeat finally imminent . . . than ever before: Ibid. See also firsthand accounts by Vera Blau and Zvi Klein about the Nazis' brutality during this Death March.

92 There was neither food nor water: Eva Mozes, Peter Somogyi, twins at Auschwitz, personal interviews.

92 With no guards . . . still left standing: Ibid.

92 On January 27, 1945 . . . Auschwitz: *Holocaust Encyclopedia*, p. 116.

93 The twins were given . . . to wear before: Eva Mozes, personal interview.

93 The jackets and trousers . . . from falling off: Ibid.

93 That night, there was a big party to celebrate the liberation: Ibid.

93 Mengele's twins . . . the adult goings-on: Ibid.

93 The scene was captured by . . . from Fascist hands: Film, courtesy Auschwitz Museum, Auschwitz, Poland.

93 The twins had to roll up . . . close-up shots: Eva Mozes, personal interview.

93 They were led out . . . looked sufficiently authentic: Ibid.

93 Menashe Lorinczi . . . their tour guide: Menashe Lorinczi, twin at Auschwitz, personal interview, March 1984.

93 Menashe took both . . . the various facilities: Ibid.
93 To his pride and delight . . . around the world: Ibid.
94 Some of the twins . . . city of Katowice: Eva and Miriam Mozes, personal interviews.
94 There were joyful rides . . . without exacting a fare: Ibid.
94 Zvi Spiegel . . . of the mass exterminations: Zvi Spiegel, personal interview.
94 He had sworn to take all the boys home: Peter Somogyi, personal series of interviews.
98 For much of her stay . . . desperately ill: Judith Yagudah, personal interview, March 1984.
98 Hoping to save her . . . powerful medications: Ibid.
98 But despite their efforts . . . to be lethal: Ibid.
98 Mrs. Rosenbaum and Judith . . . Cluj, in Romania: Ibid.
99 Some children set off alone: Hedvah and Leah Stern, personal interview, May 1988.
99 Others found adults . . . part of the journey: Eva and Miriam Mozes, personal interviews.
100 By the time Mengele's twins . . . Poland's Upper Silesia: David Marwell, OSI, personal interview.
100 There, Mengele slipped . . . of camp physician: Ibid.
100 He left Gross Rosen . . . February 11, 1945: Ibid.
100 However, he was spotted by . . . Czechoslovakia: Samuel and Mordecai Bash, twins at Auschwitz, personal interview, Tel Aviv, Israel, March 1984.
101 At least two claim . . . immediately: Ibid.
102 Mengele quickly shed . . . of the Wehrmacht: *Bunte* II.
102 He joined . . . the defeated army: Ibid.
102 At last, the unit surrendered . . . a prisoner of war: Ibid.
103 Confined to a POW camp . . . account of that period: David Marwell, OSI, personal interview.
103 By June 1945, American Occupation forces . . . jails and internment camps: Robert Wolfe, director of Captured German Records, U.S. National Archives, Washington, D.C., August 1990.
103 Since all SS officers . . . sign that it had been removed: Ibid.
103 Mengele's supreme vanity . . . to the tattoo: David Marwell, OSI, personal interview.
103 On August 18, 1945 . . . in the SS: Ibid.
103 But U.S. Occupation forces . . . wrote years after the war: *Bunte* II, p. 3.
104 Although purportedly a fictional . . . after leaving Auschwitz: Neal Sher, director, and David Marwell, historian, OSI, personal interviews.
104 When Mengele arrived . . . in need of a job: *Bunte* II, p. 4.
104 In an interview . . . could do the job: Ibid.
104 He even had time . . . favorite activity, reading: Ibid.
104 According to the Fischer family . . . called upon to entertain: Ibid.
104 The only member . . . the farmer's brother, Alois: Ibid.
105 Alois concluded that he . . . house to hide out: Ibid.
105 The war criminal . . . for over three years: Marwell, personal interview.

106 Anxious for news about . . . to visit Günzburg: Mengele autobiography excerpts, courtesy of OSI.
106 He let everyone know their beloved Beppo was safe: Ibid.
107 Mengele's father was being scrutinized . . . detention camp: OSI archives.
107 Lolo, Josef's brother . . . a camp in Yugoslavia: *Bunte* III, p. 8.
107 As for Irene . . . by the Mengele family: *Bunte* II, p. 5.
107 Irene's parents accompanied her . . . in nearby Autenried: "Researchers Find Signs U.S. Knew of Mengele's Postwar Hideout," *Washington Post*, March 15, 1985.
107 Specialists . . . other units and agencies: Marwell, personal interview.
107 As survivors emerged . . . more and more frequently: Mrs. Julia Kane, former Günzburg resident, personal interview.

5 The Trial That Never Was

110 Irene Shoenbein Mengele cut . . . her black mourning clothes: Mrs. Julia Kane, personal interview.
110 Contemporaries recall how well . . . mournful she looked: Ibid.
110 But Autenried and Günzburg . . . famine or no famine: *Bunte* II, p. 4.
110 Although the Mengele family . . . missing and apparently dead: Papers filed by attorneys for Karl Mengele, Sr., in an effort to free him from detention, in possession of OSI.
110 By a strange twist . . . the Mengele mansion: David Marwell, OSI, personal interview.
110 The family rushed over . . . lying around the house: Ibid.
110 An investigation by . . . Mengele clan in Günzburg: Ibid.
110 The Americans stationed . . . the Justice Department probe: Ibid.
111 The U.S. commander . . . help put it out: Ibid.
111 In fairness to . . . immediately after the war: Benjamin Ferencz, former U.S. Chief Counsel for War Crimes, personal interview, June 1985.
111 The British, Americans, Poles . . . for the war machine: Robert Wolfe, director of Captured German Records, U.S. National Archives, July 1990.
111 But even compared . . . death-camp doctor seemed a minor player: Ibid.
111 The Office . . . for wanted Nazis: Benjamin Ferencz, personal interview.
111 Because of the need . . . nature to the manhunts: Ibid.
111 The teams were also . . . slipping out of the country: Robert Wolfe, personal interview, July 1990.
111 As Benjamin Ferencz . . . the emphasis was on speed: Benjamin Ferencz, personal interview.
111 Once a Nazi . . . for a trial: Ibid.
111 In the course . . . crimes against humanity: Robert Wolfe, personal interview, July 1990.
111 Three others received life sentences . . . acquitted: Ibid.
112 Others, like Hermann Göring committed suicide rather than face the public reckoning: Ibid.
112 Mengele, safely ensconced . . . with unvarnished contempt: David Marwell, OSI, personal interview.

112 In his autobiographical novel . . . to "political theater": Mengele autobiography, courtesy of OSI.
112 Nuremberg, says Andreas . . . brought to its knees: Ibid.
112 The former death-camp doctor . . . except to lose the war: Ibid.
112 During one of his . . . "before him had done": Ibid.
112 The former commandant . . . experiments on twins: Posner and Ware, *Complete Story*, p. 76.
112 A former Auschwitz inmate, Marie-Vaillant Claude Couturier, described . . . interest in twins: Testimony of Marie-Vaillant Claude Couturier, International Military Tribunal, January 28, 1946. See Vol. VI, p. 240 (German Language Edition), Trial of Major War Criminals, IMT, in possession of U.S. National Archives.
113 A perusal of Mengele's writings . . . a steadfast Nazi: Professor Uwe-Dietrich Adam, German historian and author who was permitted to review Mengele's papers, personal interview, Frankfurt, May 1985.
113 Bitter and angry over . . . American occupation forces: Mengele autobiography, courtesy of OSI.
113 Mengele felt the Jews . . . they later encountered: Ibid.
113 In his novel . . . about the Nazis: Posner and Ware, *Complete Story*, p. 80.
113 "Hitler warned Jews" . . . tells a friend: Ibid.
113 "This fact alone . . . the concentration camps": Ibid.
113 As "potential enemies" . . . had to be killed: Mengele autobiography, courtesy of OSI, translation by David Marwell.
115 As the international military tribunal . . . more and more countries: Robert Wolfe, U.S. National Archives, September 12, 1990.
115 Held in Lüneberg, site of the old Bergen-Belsen concentration camp, these trials . . . was getting under way: Ibid.
116 Many of the Lüneberg defendants had served both at Auschwitz and at Bergen-Belsen . . . including some SS doctors: Excerpts from the Belsen trials, Reference Files, U.S. National Archives.
116 One of them, Dr. Josef Kramer . . . physician at Auschwitz-Birkenau: Ibid.
116 According to former inmates . . . that Dr. Mengele feared: Olga Lengyel, *Five Chimneys*, p. 147. Mengele, "so sure of himself before impotent internees, trembled in the face of the 'beast of Belsen,' " Lengyel wrote.
116 After the fall of Auschwitz, Irma had sought to continue her death-camp career at Bergen-Belsen: Reference Files, U.S. National Archives.
116 She would beat prisoners . . . they fell unconscious and died: Ibid.
116 Only at the end . . . with a semblance of kindness: Ibid.
116 During the trial . . . savagery and cruelty: Ibid.
116 In her own defense . . . most notably, Dr. Josef Mengele: Ibid.
116 She had "always" accompanied Mengele; he had been her boss: Ibid.
116 Mengele's fair companion was hanged, along with Dr. Kramer: Ibid.
116 By the end of the Lüneberg trial, much was already known about his actions as an SS camp doctor: Dr. Robert Wolfe, personal interview.
117 As new, exhaustive manhunts . . . under his alias of "Fritz Hollmann": David Marwell, personal interview.

117 And whenever he could . . . and even talked about Mengele's future hopes: Mengele autobiography, courtesy of OSI.
118 Mengele's cunning old patron . . . his former assistant: Dr. Klaus-Dieter Thomann, *Frankfurter Rundschau*, p. 8.
118 In February 1945 . . . entire library with him: Report by U.S. Forces, European Theater, Military Intelligence Service Center, Document Control Section—APO 757, December 11, 1946, U.S. National Archives.
118 This included eighty-eight cases . . . small inns around Solz: Ibid.
118 It is also believed . . . Mengele at Auschwitz: Dr. Klaus-Dieter Thomann, personal interview, May 1985.
118 He claimed to the press . . . and mass murders: Wolfson report, pp. 4, 6.
118 Verschuer said he assumed . . . "of natural causes": Ibid.
118 In order to make sure . . . excesses onto his protégé: Dr. Klaus-Dieter Thomann, personal interview, May 1985.
118 Just how Mengele felt at seeing himself blamed . . . in any of his papers: Professor Uwe-Dietrich Adam, personal interview, May 1985.
118 For example, he told . . . a vaccine for tuberculosis: Dr. Klaus-Dieter Thomann, personal interview.
118 It was an ingenious idea . . . a vaccine or a cure: Ibid.
118 If the Denazification authorities . . . trifling sum of six hundred marks: Ibid.
119 Now, Verschuer knew exactly . . . trust of the postwar establishment: Wolfson report, p. 4.
119 They pretended the Auschwitz doctor was aberration who had distorted the "ideals" of eugenics: Dr. Klaus-Dieter Thomann, personal interview. See also Müller-Hill, *Todliche Wissenschaft*, p. 83.
119 Two of his more ethical former colleagues . . . during the war: Wolfson report, p. 2. The outspoken scientists were Dr. Robert Havemann and Dr. K. Gottshaldt.
119 The two scientists . . . going on at Auschwitz: Ibid.
119 They charged he had known about . . . Jews put to death by Mengele: Ibid., pp. 3–4, 6.
119 As an intimate . . . he himself had visited Auschwitz: Ibid.
120 The scandal that resulted . . . career in Frankfurt: Thomann, *Frankfurter Rundschau*, May 21, 1985, p. 8.
120 More significantly, the ensuing brouhaha . . . authorities at Nuremberg: Wolfson report, pp. 1–7.
120 At the Berlin office . . . affair of Doctors Mengele and Verschuer: Dr. Manfred Wolfson, personal interview, undated.
120 He interviewed their former colleagues . . . more notorious Auschwitz doctor: Wolfson report, pp. 1–7.
120 As Wolfson probed Verschuer's past . . . his ties with the Nazis: Ibid., pp. 3–4.
120 He also noted . . . thirties and early forties: Ibid., p. 4.
120 In a 1946 report . . . under the aegis of Verschuer: Ibid., pp. 1–7.
120 Wolfson wrote that . . . traced back to Verschuer: Ibid., pp. 3, 6–7.
120 His old professor . . . or "heterochromatic" eyes: Ibid.
120 To satisfy Verschuer's needs . . . to him in Berlin: Ibid., p. 3.

120 These gruesome discoveries prompted Wolfson to recommend that Verschuer be "interrogated and tried": Ibid., p. 7.
120 He cited witnesses . . . tests on inmates: Ibid., pp. 5–6.
120 He noted . . . sterilization of women: Ibid., p. 6.
120 "Twins and triplets" . . . one survivor had told Wolfson: Ibid., p. 5.
120 This man recounted . . . Auschwitz in July 1944: Ibid.
121 "Of course!" Mengele reportedly replied, cheerful as ever: Ibid., p. 5.
121 But as the grieving father . . . "nor had any news": Ibid., p. 5.
122 At the conclusion . . . "indicted for war crimes": Ibid., p. 7.
122 Mengele's homesickness was relieved . . . Karl in October 1946: Mengele autobiography, courtesy of OSI, translation by David Marwell.
122 Karl confirmed the sad news of their mother's death and their father's imprisonment: Ibid.
122 The two men . . . business was faring: Ibid.
122 Years later, Mengele relived . . . tells Andreas: Ibid.
123 Flashing her nicest smile . . . and was presumed dead: *Bunte* II, p. 3.
123 While the officer . . . had been more skeptical: Ibid., p. 3.
123 The two men had . . . nothing of her husband's whereabouts: Ibid.
123 Since their marriage . . . spent little time together: *Bunte* II, p. 4.
123 But now, peace was here at last . . . still they were apart: *Bunte* II, p. 5.
123 As their son, Rolf . . . family life she craved: Ibid.
126 In the fall of 1946 . . . medical establishment were indicted: Jack S. Boozer, "Children of Hippocrates: Doctors in Nazi Germany," *Annals of the American Academy of Political and Social Science*, Vol. 450 (July 1980), p. 84.
126 It is ironic that this trial, where Mengele ought to have been the star defendant, came and went without him: Ibid., p. 90.
126 The Wolfson memorandum . . . urging that they be indicted: Wolfson report, p. 7.
126 What became of it . . . settled on the West Coast: Manfred Wolfson, personal interview, 1986. Wolfson is still passionate on the subject of Mengele and Verschuer. In a letter dated June 5, 1986, addressed to Susan Leon, the editor of this book, he writes that "in a sense, von Verschuer was more important than Mengele, because while Mengele was known to the Auschwitz inmates, he was an implementor of policy . . . Verschuer's writings in Nazi academic journals were widely circulated and argued for legitimacy on race and genetic grounds for the Nazi policy of genocide."
126 He engaged in far greater atrocities . . . example of death-camp physician: Dr. John Mendelsohn, deceased, former archivist in charge of war crimes records, U.S. National Archives, series of interviews, May–July 1985, Washington, D.C.
126 Hoven, chief doctor of Buchenwald . . . conducted medical experiments: Ibid.
126 Conversely, those doctors . . . for the gas chambers: Ibid.
127 "If they could have" . . . until his death in 1986: Ibid.
127 Mendelsohn believed that Mengele's case . . . the international military tribunals: Ibid.

127 The proceedings revealed that over two hundred . . . had stood silently by: Alexander Mitscherlich and Fred Mielke, *Doctors of Infamy: The Story of the Nazi Medical Crimes* (New York: Henry Schuman, 1949), x.
127 At the trial, female witnesses tearfully . . . at Auschwitz and Ravensbruck: Boozer, *Children of Hippocrates*, pp. 85–87.
127 But oddly enough, no witness . . . the Angel of Death, Dr. Mengele: Dr. Robert Wolfe, personal interview, July 1990.
127 According to Neal Sher . . . on the prosecutors' lists: Neal Sher, director, Office of Special Investigations, U.S. Justice Department, July 10, 1990.
127 Another Holocaust expert . . . "through our fingers": Dr. Robert Wolfe, personal interview, July 1990.
128 But in defense of . . . "caught as many as we did": Ibid.
128 It's remarkable we even had a doctors' trial because people were getting tired: Ibid.
128 Sure enough, by late 1946 and 1947 . . . in the U.S. Congress: Ibid.
128 If fighting them meant joining forces . . . the United States was ready to do so: Erhard Dabringhaus, *Klaus Barbie: The Shocking Story of How the U.S. Used This Nazi War Criminal as an Intelligence Agent* (Washington, D.C.: Acropolis Books, 1984), pp. 48, 51–53, 64–65.
128 American funding for Nazi prosecutions . . . fight against Communism: Dr. Robert Wolfe, personal interview, September 1990.
129 In Augsburg, near Günzburg . . . and operative in April 1947: Dabringhaus, *Barbie*, pp. 52–53.
129 Yet, instead of being tried . . . infiltration of Germany: Ibid., p. 78. Dabringhaus, the U.S. agent who worked with Barbie for over a year, notes that Barbie's benefits as an American spy included "free housing, supplies and money, which added up to about $2000 a month overall."
129 It is probable, however . . . a useful contact: Dr. Klaus-Dieter Thomann, personal interview.
129 Years later, Manfred Wolfson . . . pretrial report, dropped: Manfred Wolfson, personal interview.
129 In 1947, Gisella Perl . . . of life at Auschwitz: Perl, *Doctor in Auschwitz*, op. cit.
129 In graphic, vivid prose . . . because she was so lovely: Ibid., p. 138.
130 At last, here was the chance to denounce the monster: Letter, Gisella Perl to Civil Affairs Division, January 27, 1947, RG 165, Records of the War Department, U.S. National Archives.
130 In her letter . . . murderer of the twentieth century: Ibid.
130 She wrote to them yet again in the fall of 1947, renewing her offer: Letter, Gisella Perl to Colonel Damon Gunn, Civil Affairs Division, October 7, 1947. RG 165, Records of the War Department, U.S. National Archives.
130 A review of the U.S. Army's file . . . "Hold for Final Action": Ibid.
130 A small note . . . erroneously that he had been "tried by Poles": Ibid.
130 By the time of her second letter, they wrongly assumed . . . Polish government: What clearly added to the confusion was the fact that Colonel Gunn left his post in the War Crimes Office. He wrote Perl a letter on October 14, 1947, informing her that he was no longer handling Nuremberg trial matters, and

reassuring her that her offer had been referred to his colleagues. Letter from Colonel Damon Gunn to Gisella Perl, U.S. National Archives.

131 Perl's October 1947 letter . . . bureaucrat to bureaucrat in Washington: Ibid.

131 By December, it had finally . . . lead prosecutor at Nuremberg: Memo from Colonel Edward H. Young, chief of War Crimes Branch, to U.S. Chief of Counsel (for War Crimes, General Telford Taylor), December 8, 1947. The memo reveals that Young in Washington did try to verify whether Mengele had in fact been tried by the Poles. It specifically requests Taylor's office to find out what happened and to advise his division "of the results of the trial by the Poles of the Auschwitz Concentration Camp Case, and particularly of the disposition of the case against Dr. Mengelerle [sic], in order that a reply be made to Dr. Perl." The file does not indicate whether Young received a reply correcting the file's earlier misinformation that Mengele had been "tried by Poles."

131 In January 1948, General Taylor . . . delay in responding: Letter from Telford Taylor to Colonel Edward Young, January 19, 1948, U.S. National Archives.

131 Taylor's brief letter ended . . . "Mengerle [sic] is dead as of October 1946": Ibid.

131 Even an exhaustive Justice Department inquiry more than forty years later failed to clarify what happened: David Marwell, personal interview.

131 Beginning with Irene's clever donning of mourning clothes . . . succeeded: Ibid.

131 The only other possible conclusion . . . such a theory: Ibid.

131 His writings on the period . . . from his unhappy wife: Mengele autobiography, courtesy of OSI.

131 He remained deeply upset over his crumbling marriage: Ibid.

131 He continued to mourn his mother's passing and the Günzberg life he would never again know: Ibid.

6 The Story of Andreas

134 Three years after . . . at the Fischer farm: Mengele autobiography, courtesy of OSI.

134 But as the terrible decade . . . insulated Bavarian town: Inge Byhan, *Bunte* II, p. 4.

134 Lights were turned on at night for the first time: Ibid.

134 They continued to meet . . . smoothed over by temporary reconciliations: David Marwell, OSI, personal interview.

134 He could also be jealous . . . inventing lovers and trysts: Mengele autobiography, courtesy of OSI, translation by David Marwell.

134 Again and again . . . relations with other men: Ibid.

135 It is by way . . . relationship with Irene: David Marwell, personal interview.

135 While hiding behind . . . a lover, and a man: Ibid.

135 In the novel, Andreas . . . to a Bavarian resort: Mengele autobiography, courtesy of OSI, translation by David Marwell.

135 Crazed with jealousy . . . "of my crumbling life": Ibid.

135 After a long, tiresome train journey . . . Irmgard is staying: Ibid.

135 She has already left . . . with her lover: Ibid.
135 "I am not an old soldier's wife . . . back home again": Ibid.
135 "But you cannot come home," Irene's alter ego, Irmgard, says at one point": Ibid.
135 Although Andreas and Irmgard . . . "disappointing": Ibid.
135 Mengele the novelist . . . "the marriage was over": Ibid.
135 Elsewhere, Andreas sadly observes . . . "my marriage also ended": Ibid.
136 He had heard that . . . especially Argentina: Byhan, *Bunte* II, p. 5
138 Although his marriage . . . to South America: Ibid.
138 "There are people" . . . the fictionalized memoirs: Mengele autobiography, courtesy of OSI.
138 "We must get through the present and start over abroad": Ibid.
138 But Irene was . . . was a mass murderer: Byhan, *Bunte* II, p. 5.
139 She categorically refused . . . her native Freiburg: Ibid.
139 "Man, woman, and child" . . . he sternly observes: Mengele autobiography, courtesy of OSI.
139 In 1948, Mengele returned . . . plan his getaway: Posner and Ware, *Complete Story*, p. 87.
139 But instead of . . . near the old town: Ibid.
140 Irene finally managed . . . it was quickly discarded: Byhan, *Bunte* II, p. 6.
140 Mengele had no choice . . . without any papers: Mengele autobiography, courtesy of OSI.
140 Mengele hired several guides to help with his escape: David Marwell, OSI, personal interview.
140 He began by taking a train to Innsbruck: Ibid.
140 He reached the border . . . on Easter Sunday, 1949: Inge Byhan, *Bunte* magazine 1985 series on Josef Mengele, Part III, Bunte Archives 28/85. Munich, Germany. Handwritten translation by Elizabeth Bergmann.
140 A guide helped him . . . dead of night: Ibid.
141 In Mengele's thinly disguised account . . . crossing: Mengele autobiography, courtesy of OSI.
141 The hero, somewhat mortified . . . "top physical shape" Mengele autobiography, courtesy of OSI.
141 Mengele's novel provides . . . dotting the Alps: Inge Byhan, *Bunte* III, p. 3.
141 Once Andreas is . . . fend for himself: Ibid.
141 He continues walking until he reaches Italy, and boards a train to Sterzing . . . to meet him: Ibid.
141 But everything goes . . . to meet him: Ibid.
141 The fugitive spends . . . Golden Cross Inn: David Marwell, OSI, personal interview.
141 Yet another contact . . . forged German ID card: Inge Byhan, *Bunte* III, p. 3.
141 In writing the "novel" . . . place names and individuals: David Marwell, personal interview. It was Marwell, through the meticulous investigation he conducted for the U.S. Justice Department, who was able to figure out the "real life" names and places alluded to in Mengele's autobiography. He was kind enough to share his findings with the author, in a series of interviews.
141 Historians who have tracked the war criminal's path . . . Taormino: Ibid.

141 The document was made out to "Helmut Gregor," a pseudonym Mengele would use for years to come: Ibid.
141 When Italy fell . . . their own identity cards: Ibid.
141 But after the war . . . dispute over sovereignty: Ibid.
142 Thanks to this card . . . safely and legally: Ibid.
142 Before Mengele left Sterzing . . . from the family: Byhan, *Bunte* III, pp. 3–4.
142 Karl Mengele, who had finally . . . and months ahead: Ibid.
143 A benevolent, even humanitarian. . . . in Karl Mengele: Hermann Abmayr, personal interview.
143 For not only . . . favorite and most generous philanthropist: Ibid.
143 He got into the habit of placing large sausages in the windows of the homes of Günzburg's impoverished residents: David Marwell, personal interview.
143 But Karl Mengele clearly possessed a munificent spirit: Dr. Kurt Lambertz, personal interview.
143 For their Italian reunion . . . sent home for safekeeping: Byhan, *Bunte* III, pp. 3–4.
143 Carrying the money . . . "Kurt" in the novel: Ibid., p. 4.
143 Plump but agile . . . and prima donna attitude: Ibid.
144 "You are not exactly a harmless tourist," he snaps at his erudite charge: Mengele autobiography, courtesy of OSI.
144 The ID card . . . war refugees legal passports: David Marwell, personal interview.
144 Using this route . . . to flee Europe: Ibid.
144 At the Swiss consulate . . . Did she recognize him? Byhan, *Bunte* II, p. 4.
144 His worst suspicions . . . issued him was useless: Ibid., p. 5.
145 A Croatian doctor . . . administered the required vaccines: Ibid.
145 Mengele had to undergo another physical exam at the port of Genoa, where his ship bound for Argentina was docked: Ibid.
145 There, Mengele couldn't help noting . . . of transmitting infection: Ibid.
145 Mengele's Genoan contact . . . exchanged money for his favors: Mengele autobiography, courtesy of OSI.
145 In the novel, the vessel . . . the *North King*: David Marwell, personal interview.
146 In the novel . . . exit visas twenty thousand lire: Byhan, *Bunte* II, p. 5.
146 In real life, Josef Mengele . . . for his murderous deeds: David Marwell, OSI, personal interview.
146 He paced up and down . . . according to his novel: Byhan, *Bunte* II, p. 6.
146 In an ironic turn . . . in former years: Ibid.
146 There was a dwarflike, handicapped . . . a morphine addict: Ibid.
146 But at last . . . returned from his holiday: Ibid.
147 Mengele's escape . . . Angel of Death had eluded capture: David Marwell, personal interview.
147 There was no sophisticated Nazi network in operation to ferret him out to safety: Ibid.
147 He was not aided by the American government or any of its intelligence agencies: Neal Sher, director, OSI, personal interview.
147 The Vatican had no apparent role in his flight: David Marwell, personal interview.

147 In chaotic postwar Europe . . . for a fee: Ibid.

147 His German credentials were not enough . . . possibly for many years: Mengele autobiography, courtesy OSI.

7 FUGITIVE'S IDYLL

149 Of course, Evita . . . threat to Perón's power: Joseph A. Page, *Perón: A Biography* (New York: Random House, 1983), p. 85.

149 As for the Nazis . . . of his humanitarian bent: Ibid., p. 88.

150 They knew that Perón would not tolerate any displays of anti-Semitism: Ibid., p. 90.

150 Similarly, thousands of Germans . . . those available in Europe: *Ronald C. Newton, German Buenos Aires, 1900–1933, Social Change and Cultural Crisis* (Austin: University of Texas Press), xiv.

150 They founded German clubs . . . in the New World: Newton, *German Buenos Aires*, p. 27.

150 And there were also several Nazi . . . years after the war was over: Newton, *German Buenos Aires*, p. 170.

150 Endemic to the community . . . that underscored every activity: Ibid., p. 172.

151 But as long as Juan Perón . . . of Argentina's upper crust: Joseph Page, *Perón*, p. 90.

152 He arrived in Argentina penniless . . . during his Genoa adventures: Inge Byhan, *Bunte* IV, p. 7.

152 Mengele was forced . . . each month from Europe: Ibid., p. 8.

153 He made do . . . end of the hall: Ibid.

153 In actuality, he . . . soldiers than its women: Ibid.

153 In letters he composed . . . "cowardly inferiority complexes": Ibid.

153 He admired the respect . . . of the German Army: Ibid.

153 Mengele went to see him with high hopes: Ibid., p. 7.

153 The encounter was disappointing . . . employed in a weaving mill: Ibid.

154 The carpenter was quitting his job . . . to have it: Ibid., p. 8.

154 The Auschwitz doctor . . . with an engineer: Ibid.

155 Then, when his roommate's daughter became sick, Mengele was asked to treat her: Ibid.

155 Although the little girl lived elsewhere with her mother, her anxious father brought her to Mengele for expert medical care: Ibid.

156 Mengele began corresponding . . . his six-year-old son: David Marwell, personal interview.

156 In 1948, she had become friendly . . . and the two fell in love: Posner and Ware, *The Complete Story*, p. 108.

156 Hackenjos, a prosperous . . . stability she craved: Herman Abmayer, personal interview.

157 Mengele's letters to Rolf . . . always signed "Uncle Fritz": Inge Byhan, *Bunte* IV, p. 8.

157 He spun colorful tales . . . cattle on the pampas: Inge Byhan, *Bunte* 1, p. 4.

157 Sometimes there were amusing drawings . . . especially for Rolf: Ibid.

157 The little boy waited expectantly . . . the father he had never known: David Marwell, personal interview.
158 Mengele faithfully kept up . . . Karl Jr. in 1949 at the age of thirty-seven, after an illness: Inge Byhan, "A Family History," *Bunte* III, p. 2.
158 Josef's father and Lolo . . . their right-hand man: Professor Zdenek Zofka, personal interview, September 1990.
158 They helped him financially as much as they could: Ibid.
158 His father even sent . . . to start a business: Inge Byhan, *Bunte* III, p. 8.
158 Josef began selling . . . was a lucrative market: Posner and Ware, *Complete Story*, p. 105.
159 One of these was Wilhelm Sassens . . . Nazi publications in Argentina: Inge Byhan, *Bunte* III, p. 8.
159 It was Sassens . . . Colonel Hans Ulrich Rudel: Ibid.
159 During World War II, Argentina . . . to the Western Hemisphere: Page, *Perón*, see pp. 87–91.
159 According to Ronald Newton . . . "to the surrounding culture": Newton, p. xv.
159 Pro-Nazi publications flourished . . . of the German Army: Arnold Forster, "The Argentine Peril," 1962 draft memorandum for Anti-Defamation League of B'Nai B'Rith, pp. 3–4. Property of ADL archives, New York City.
159 One of them, the virulently anti-Semitic *Der Weg*, or "The Way," was popular in the early 1950s: Ibid.
159 In 1953, an article appeared . . . Mengele's alias, "Helmut Gregor": G. Helmuth, "*Die Vererbung Als Biologischer Vorgang*," *Der Weg*, pp. 815–820, undated. Copy of article found in Mengele files obtained by *Stern* magazine. The Justice Department's Office of Special Investigations provided author with the year of publication.
160 On the other side . . . of his old pupil by 1953: Klaus-Dieter Thomann, personal interview.
161 After maintaining a low profile . . . the University of Münster in 1951: Thomann, *Frankfurter Rundschau*, p. 8.
161 Once installed at Münster . . . studies in West Germany: Ibid.
161 Verschuer was careful to omit . . . lectures or publications: Thomann, personal interview.
161 He served for many years . . . of Nazi racial science: Roger Pearson, editor, *Mankind Quarterly*, personal interview, July 10, 1990.
161 His former deputy . . . the status of Verschuer: Thomann, *Frankfurter Rundschau*, p. 8.
161 A diehard racial scientist long after . . . "inferior" genetic stock: Ibid.
161 Schade became an adviser . . . to neo-Nazi movements: Ibid.
161 He and Verschuer were on the best of terms: Ibid.
161 According to Hans Sedlmeir . . . side trip to see Verschuer: David Marwell, personal interview.
162 There is indeed evidence . . . the furor surrounding his pupil's name: Dr. Klaus-Dieter Thomann, personal interview.
162 Mengele also apparently saw a need . . . not one mentions Verschuer: Uwe-Dietrich Adam, personal interview.
162 There are hints . . . but few details are available: Ladislas Farago, *Aftermath:*

Martin Bormann and the Fourth Reich (New York: Simon and Schuster, 1974), p. 275.

162 In 1954, the Auschwitz doctor . . . many affluent Germans lived: Ibid.

162 Shortly thereafter, he purchased another . . . same elegant neighborhood: Ibid.

162 It was right near Perón's old estate: Inge Byhan, *Bunte* IV, p. 5.

163 He was overjoyed when . . . sporty new Borgward "Isabella": Ibid.

163 In 1954, Mengele's father . . . visit his son: Ibid.

164 The couple was divorced that year: Ralph Blumenthal, "Mengele Family Members Are United in Their Silence," *New York Times*, June 11, 1985, p. 2.

164 But when Juan Perón was ousted . . . He had protected both: Arnold Forster, *Argentine Peril*, p. 4.

164 In fact, many former Nazis were flourishing in West German society: Hannah Arendt, *Eichmann in Jerusalem: A Report on the Banality of Evil* (New York: Viking Press, 1963), pp. 17–19.

164 Some of the worst offenders of the Third Reich were leading comfortable lives, and, indeed, had achieved positions of prominence in the "new" Germany: Ibid.

165 Throughout Latin America, war criminals . . . of the Third Reich: Eliot Welles, Anti-Defamation League of B'Nai B'Rith, personal interview, September 10, 1990.

166 Sometime that year . . . from Helmut Gregor to Josef Mengele: Rabbi Morton Rosenthal, director, Latin American Affairs, B'Nai B'Rith Anti-Defamation League, *Testimony Before the U.S. Senate Judiciary Subcommittee on Juvenile Justice*, March 19, 1985, p. 3.

166 Years later, when the hunt . . . how the Angel of Death of Auschwitz really looked: Neal Sher, OSI, personal interview, May 1985.

167 Rolf was now a handsome lad of twelve . . . a living father: Inge Byhan, *Bunte* IV, p. 8.

167 It was the trusted Sedlmeier . . . logistics: Ibid.

167 He personally went . . . a special winter holiday: Ibid.

167 Standing next to the two . . . than in his photographs: Ibid., pp. 8–9.

167 The hotel that was selected . . . herself had once stayed: Ibid., p. 9.

167 For the first time ever . . . compliments of Uncle Fritz: Ibid.

167 At lavish meals . . . a grown-up Rolf would later admit: Ibid.

168 He suggested he accepted the veracity of the stories about Dr. Mengele: Inge Byhan, *Bunte* II, p. 2.

168 When asked whether he felt his father . . . "If he was guilty": Ibid., p. 2.

169 Rather, Rolf released old family pictures . . . and beaming as he stood by Aunt Martha: *Bunte* II, pp. 16–20.

169 In other, older photographs . . . and an overcoat: Ibid., p. 20.

169 Rolf did say . . . had never killed anyone "personally": Ibid., p. 2.

170 Some of the Germans . . . manned by slave laborers: Eli Rosenbaum, deputy director, OSI, personal interview, September 1990.

170 But their pasts were swept aside . . . valued citizens of this country: Ibid.

8 The Angel Retreats

173 During her Günzburg youth . . . and beautiful face: Mrs. Julia Kane, personal interview, Westchester, N.Y.
173 At thirty-five, she took pains . . . "a fashion model": Ibid.
173 The couple was urged . . . particularly Karl Sr.: Hermann Abmayer, personal interview.
174 In 1958, Martha and Josef . . . for the ceremony: Ralph Blumenthal, "Mengele Family Members Are United In Their Silence," *New York Times*, June 11, 1985, p. 3.
174 Josef's boyhood friends . . . his late brother's wife: Dr. Hermann Lieb, personal interview, May 1985.
174 Those who knew Martha were taken aback . . . South American backwater: Mrs. Julia Kane, personal interview.
174 In Günzburg, Martha . . . gold digger: Ibid.
174 Years earlier, she had become pregnant . . . named Wilhelm Ensmann: Serge Klarsfeld, Nazi-hunter, personal files, obtained May 1985, Paris, France.
174 She promptly divorced Ensmann and married Karl: Ibid.
174 The baby, Karl Heinz . . . he was Karl's child: Ibid.
174 It was hard to imagine . . . the comforts of Europe: Mrs. Julia Kane, personal interview.
176 Within months of their marriage . . . by the local police: Ladislas Farago, *Aftermath*, pp. 274–75.
176 The authorities were investigating . . . without a license: Inge Byhan, *Bunte* IV, p. 8.
176 He was questioned . . . charges were quickly dropped: "The Mengele Mystery," *Time*, June 24, 1985, p. 42.
176 Mengele's life was marred . . . father died in 1959: Ralph Blumenthal, "Mengele Family Members," *New York Times*, June 11, 1985, p. 3.
176 Hundreds of people came . . . quick, furtive trip to attend his father's funeral: Paul Lewis, "Couple Will Face a German Inquiry on Aid to Mengele," *New York Times*, June 10, 1985.
176 "Although the reports are contradictory . . . to pay his final respects: Ibid., testimony of Wolfram and Liselotte Bossert before West German Prosecutor's Office.
176 That same year, 1959 . . . to locate the Auschwitz doctor: Simon Wiesenthal, *The Murderers Among Us* (New York: McGraw-Hill, 1967), p. 156.
176 It was spurred by . . . who had escaped punishment: Ibid.
176 In Vienna, an Auschwitz . . . of the selection line: Ibid.
177 Ironically, it was Mengele's divorce . . . his South American retreat: Ibid.
177 Irene and Josef's divorce papers . . . address as Buenos Aires: Ibid.
177 By uncovering these documents . . . in Argentina under his own name: Ibid.
177 Langbein set about amassing . . . extradited and tried: Ibid.
178 Germany issued its first arrest warrant for Mengele on June 7, 1959: Ladislas Farago, *Aftermath*, p. 277.
178 He left for Paraguay to seek refuge with his friends in the German community there: Posner and Ware, *The Complete Story*, p. 125.

181 When Mengele applied . . . and by their mutual friend Hans Ulric Rudel: Ladislas Farago, *Aftermath*, p. 276.
181 He was now "José Mengele": Citizenship papers of José Mengele, November 27, 1959, courtesy of government of Paraguay.
181 Newly declassified files . . . to help out the Angel of Death: Foreign Service Dispatch, "Extradition Case of Nazi Josef Mengele, Accused of War Crimes," from American Embassy in Buenos Aires to State Department, June 24, 1960.
181 According to a cable sent . . . Mengele to be extradited: Ibid.
181 The Argentines coolly replied . . . "entry into this country": Ibid.
181 With remarkable gall . . . allegations against Mengele: Ibid.
181 The Germans didn't respond . . . address to Argentina: Ibid.
182 When they did . . . accede to the Germans' request: Ibid.
182 Instead, Buenos Aires officials . . . "for a recommendation": Ibid.
182 "But by then, Mengele" . . . read the cable from the U.S. embassy: Ibid.
182 According to an Israeli diplomat . . . luxurious villa in Olivos: Confidential interview with a former Israeli envoy to Buenos Aires, Tel Aviv, Israel, May 1985.
183 They apparently hoped to capture Mengele at the same time: Isser Harel, *The House on Garibaldi Street* (New York: Viking Press, 1975), p. 214.
183 The agents persuaded Eichmann . . . it was too late: Harel, *Garibaldi Street*, p. 215.
183 Eichmann had been living under the pseudonym of Ricardo Klement: Ibid., p. 61.
183 The man in charge . . . a wife and three children: Ibid., p. 61.
183 Instead of residing in . . . neither running water nor electricity: Ibid., p. 62.
183 Wilhelm Sassens, the journalist . . . had also befriended Eichmann: Hausner, *Justice in Jerusalem* (New York: Harper and Row, 1966), pp. 9–12.
184 Martha's complaints about her rough life . . . talk of Günzburg: Mrs. Julia Kane, personal interview.
184 Residents of the sleepy town . . . to do housework: Ibid.
184 In his Paraguayan hideout . . . more careful than before: Professor Uwe-Dietrich Adam, personal interview, May 1985.
184 His friend Rudel . . . named Wolfgang Gerhard: "Who Helped Mengele?," *Newsweek*, June 24, 1985, p. 44.
184 Gerhard worked out . . . for the Auschwitz doctor: Ibid.
185 In the fifteen years . . . the perpetrators of the Holocaust: Hannah Arendt, *Eichmann in Jerusalem* (New York: Penguin Books, 1976), p. 14.
185 The Central Agency . . . in two years accomplished very little: Ibid.
185 Richard Baer . . . of the Eichmann "team": Ibid., pp. 14–15.
185 Gideon Hausner, the Israeli prosecutor . . . course of the trial: Hausner, *Justice in Jerusalem*, p. 436.
185 One young German wrote . . . "against humanity" Ibid.
185 He offered to work . . . to do the same: Ibid.
185 Their only question was, "Will you take us?": Ibid.
186 This indeed had been the outcome . . . end of the war: Arendt, *Eichmann in Jerusalem*, pp. 14–15.
187 Otto Bradfish, who killed fifteen thousand Jews . . . hard labor: Ibid., p. 15.

187 And Joseph Lechthaler . . . behind bars: Ibid.
187 Karl Wolff . . . nabbed and tried: Ibid.
187 He was quoted as saying . . . "Warsaw to Treblinka": Ibid.
187 Around the time of the Eichmann kidnapping . . . was really his father: David Marwell, personal interview.
187 The Mengeles told Rolf . . . several advanced degrees: Inge Byhan, *Bunte* IV, p. 4.
188 Anguished by the revelations . . . failing at school: Ibid., p. 6.
188 The tenor of his correspondence . . . he confided in these same interviews: Posner and Ware, *The Complete Story*, pp. 234–35.
188 The Eichmann case . . . Jewish communities the world over: Rabbi Morton Rosenthal, ADL, personal interview, September 1990.
189 In June 1960, Argentina presented a formal complaint to the U.N. Security Council protesting what Israel had done: Hausner, *Justice in Jerusalem*, p. 460.
189 In addition, Israel . . . for violating international law: Hausner, *Justice in Jerusalem*, p. 458.
189 There were anti-Semitic episodes around the world: Arnold Forster, *The Argentine Peril*, p. 1.
189 On June 24, 1962, a young . . . way to school: "Argentine Incidents," unsigned, undated memorandum, archives of B'Nai B'Rith Anti-Defamation League.
189 Four days later . . . and beat him savagely: Ibid.
189 Secret memoranda sent by . . . the anti-Semitic incidents: Courtesy, B'Nai B'Rith ADL archives.
189 As one 1962 memorandum . . . "come to the fore": Memorandum from central Jewish body of Argentina, courtesy, B'Nai B'Rith archives.
190 Isser Harel . . . was especially proud: Isser Harel, former Mossad chief, personal interview, May 1985, Tel Aviv.
190 From an operational . . . gone practically without a hitch: Ibid.
190 With the Eichmann operation . . . to apprehend the elusive Auschwitz doctor: Ibid.
190 Harel was especially anxious . . . than the bureaucrat Eichmann: Harel, *Garibaldi Street*, p. 214.
190 Harel himself has said . . . the possibility of great bloodshed: Harel, personal interview, May 1985.
190 In 1962, the final confrontation . . . not South America: Harel, *Garibaldi Street*, vi.
190 The Israelis learned that Egypt was starting up a rocket program with the help of Nazi scientists: Ibid.
191 In one retreat . . . secret agents who never arrived: "The Mengele Mystery," *Time*, June 24, 1985, p. 40.

9 Brazilian Hideaway

196 Preparations for the Frankfurt trials . . . Germany's judicial system: Wiesenthal, *Murderers*, p. 169.
196 Bauer, prosecutor general . . . where to find Eichmann: Harel, *Garibaldi Street*, pp. 3–9.

196 He had done so . . . a serious pursuit: Ibid., p. 8.
196 Courageous and energetic . . . to atone for their terrible past: Ibid., p. 9.
196 In a diary entry . . . "new witchhunt in Germany": Diary of Josef Mengele, May 2, 1962, courtesy of *Stern* magazine.
196 He was referring . . . for nearly two years: David Marwell, personal interview.
197 Mengele still longed for . . . "German nation under Adolf Hitler": Letter from Mengele to Hans Sedlmeier, undated, courtesy of *Stern*. Translated by Maria Guitart.
197 He noted that . . . "to suffer and feel guilty": Ibid.
197 Any people with Nazi ties . . . and, finally, the war criminals: Ibid.
197 Depending on one's classification . . . "a rehabilitated citizen": Ibid.
198 Even after all these years . . . Nazis to book: Ibid.
198 Quoting . . . "to tread on a dead lion": Ibid.
198 He was posing as a Swiss emigré named Peter Hochbichler: Alan Riding, *New York Times*, June 9, 1985, p. 16.
199 The Stammers were . . . Swiss exile: Ibid.
199 Mengele would assist . . . in managing: "The Mengele Mystery," *Time*, June 24, 1985, p. 40.
199 Mengele went to live with the Stammers in 1961, helping out with the chores on the coffee and fruit plantation in Nova Europa, 175 miles north of São Paulo: Ibid.
199 "Beginning of memoirs difficult," he says in one diary entry: Mengele diary, March 3, 1962.
199 In another . . . "still displeased with it": Mengele diary, February 5, 1962.
199 At one point, as Mengele . . . disturbed his concentration: Mengele diary, March 3, 1962. Even though Mengele was living in a remote farming village, it is not surprising that *Carnaval* celebrations—the most important and frenetic holiday in Brazil—would go on even there.
199 Deeply bothered by the revelry . . . on his memoirs: Ibid.
199 "Strong migraine" . . . January 24, 1962: Mengele diary, January 24, 1962.
199 The next day . . . and had an earache: Mengele diary, January 25, 1962.
200 On the twenty-sixth . . . a very bad dream: Mengele diary, January 26, 1962.
200 He did not describe . . . shouting: Ibid.
200 In one surprising entry . . . the splendid Brazilian summer: Mengele diary, February 7, 1962.
200 He noted the blossoming flowers outside . . . green fields: Ibid.
200 By early February . . . "reading poetry out loud": Mengele diary, February 10, 1962.
201 But then, the litany of complaints quickly resumed: Mengele diary, February 12 and 18, 1962.
201 These documents, which Rolf inherited . . . thoughts about Auschwitz: Mengele's son turned over the bulk of his father's voluminous writings to precisely one German publication, *Bunte*. But the rival *Stern* magazine was able to obtain thousands of pages of diaries, notebooks, and letters from the Bosserts in Argentina, who also had kept a stash of Mengele papers. It is significant to note that at no point did Mengele's son turn any of these writings over to the

German Prosecutor's Office, which had handled the investigation of his father for so many years. Rather, Chief Prosecutor Hans Eberhard Klein had to go a-begging to *Bunte* to obtain them.

201 But even a close reading . . . at the concentration camp: Professor Uwe-Dietrich Adam, personal interview, May 1985.

202 After perusing Camus's *The Fall* . . . it was terribly overrated: Mengele diary, January 23, 1978.

203 The Auschwitz doctor decried . . . "Nazi-haters and communists": Ibid.

203 He delivered a blistering critique of *The Fall*, perhaps Camus's greatest novel: Ibid.

203 "You have made it very difficult . . . always, always love you: Mengele diary, June 7, 1962.

203 "Very bad headache". . ., "Condition worsened": Mengele diary, February 13, 1962; March 21, 1962; March 20, 1962.

203 Even reading, his favorite pastime . . . his migraines: Mengele diary, March 21, 1962.

203 Sometime later that year . . . closer to São Paulo: *Newsweek*, June 24, 1985, p. 44.

203 From a perch atop the eight-foot tower . . . any unwanted visitors: Ibid.

204 One day on the radio . . . "We Praise You": Mengele diary, March 16, 1962.

204 Another day, he read a book . . . there as a young man: Mengele diary, March 14, 1962.

204 "A lot of old memories . . . and the people better": Ibid.

204 The Stammers came to dread and fear their erratic "houseguest": Alan Riding, "Woman Says Man Who Admitted He Was Dr. Mengele Died in 1979," *New York Times*, June 9, 1985, p. 1.

204 But instead of being grateful . . . in press interviews: *Newsweek*, June 24, 1985, p. 44. Ibid.

205 He even gave the Stammers advice . . . discipline them more often: *Time*, June 24, 1985, p. 40.

205 Alienated from the Stammers . . . Gerhard for friendship and support: *Newsweek*, June 24, 1985, p. 44.

205 He was also as fanatic . . . a Nazi as Mengele: Ibid., p. 45.

205 As a young man, Gerhard had joined the Hitler Youth . . . German Army: Ibid.

205 He named his oldest son Adolf . . . the glorious Third Reich: Ibid.

205 She once gave . . . in their original 1943 wrappers: James Markhan, "Mengele Doublė Called Fervid Nazi," *New York Times*, June 13, 1985.

205 They were "Jew soaps" . . . killed in the concentration camps: Ibid.

205 Another of the Gerhards' . . . Christmas tree every year: *Newsweek*, June 24, 1985, p. 45.

205 You always have to . . . Austrian liked to say: *New York Times*, June 13, 1985.

206 Although Gerhard had . . . realized his true identity: *Time*, June 24, 1985, p. 40.

206 Later that day . . . admitted he was the notorious Dr. Mengele of Auschwitz: *New York Times*, June 9, 1985, p. 16.

206 Afterward, Mengele refused . . . with the Stammers: *Time*, June 24, 1985, p. 40.

206 Intensely paranoid after . . . to the farm with suspicion: *New York Times*, June 6, 1985.
207 Ironically, visitors to the Stammer farm . . . charmed by him: David Marwell, OSI, personal interview.
207 It seemed to her that Josef Mengele had two sides, "one for strangers, and the other when he did not need to dissimulate": OSI archives.
207 Whenever the Stammers complained . . . the famed SS doctor: *Newsweek*, June 24, 1985, p. 44.
207 When that approach ceased to work, Gerhard resorted to threats: Ibid.
208 They quickly dispatched . . . to smooth matters over: Ibid.
208 Mengele continued to work on his memoirs: See 1962 Mengele diary, March 21, 24; April 3, 4, 6, 10, 16–20.
208 Progress was slow, perhaps . . . over the past: Mengele diary, April 24, 1962.
208 In February 1964, Ambassador Eckart Briest requested an audience with President Stroessner: Simon Wiesenthal, *The Murderers Among Us*, p. 169.
208 In a rare display . . . turn over the infamous death-camp doctor: Ibid.
208 Stroessner was so infuriated . . . a minor diplomatic crisis: Ibid.
209 At the trial, an Israeli doctor . . . because they were not identical: *Jerusalem Post*, August 18, 1964.
209 His story was confirmed by the Auschwitz pharmacist, Viktor Capesius: Ibid.
209 "I have no time now," . . . at the bewildered children: Ibid.
209 They were promptly taken away to be gassed: Ibid.
209 Fritz Bauer, the prosecutor . . . company of Martin Bormann: *Jerusalem Post*, July 10, 1964.
209 Bauer insisted that Bormann . . . alive and well in South America: Ibid.
209 Langbein hoped to persuade . . . they had once bestowed on Mengele: Confidential interviews with officials from the University of Munich, April 1986.
209 The University of Munich said it would follow Frankfurt's cue: Ibid.
212 His old mentor, Verschuer . . . honors and high praise: Klaus-Dieter Thomann, personal interview.
212 He had succeeded in . . . institutes in West Germany: *Frankfurter Rundschau*, May 20, 1985, p. 20.
212 He was especially upset . . . acknowledge his birthday: Mengele diary, March 16, 1975.
212 When he turned sixty . . . a birthday card: Mengele letter to Hans Sedlmeier, undated, copy courtesy of *Stern*.
212 In the postwar years . . . build up the factory: Mrs. Julia Kane, personal interview.
212 In the process . . . fair, and honorable man: Ibid.
212 A review of Mengele's letters to Alois . . . money to spare: Mengele correspondence, letter from Josef Mengele to Alois Mengele, undated. Also undated letter to Sedlmeier.
213 But the townspeople . . . sadism at Auschwitz: Interview with Rudolph Koeppler, mayor of Günzburg, May 1985.
213 According to the mayor of Günzburg . . . brother's version of events: Ibid.
214 When Alois developed cancer . . . mend fences with him: Mengele letter to Alois, undated.

214 He complained he had only learned . . . and by chance: Ibid.
214 "Perhaps that is also part . . . with you, dear brother": Ibid.
214 "As a reward" . . . last letters to Lolo: Ibid.
214 It is a great honor . . . exemplary: Ibid.
214 "You should train your offspring" . . . he warned: Ibid.
215 He stressed how proud . . . "you so greatly increased": Ibid.
215 "I am especially happy" . . . he sadly noted: Ibid.
215 Gerhard introduced him . . . also die-hard Nazis: *Newsweek*, June 24, 1985, p. 44.
216 The Bossert children "adored" Mengele . . . about Mengele's years in hiding: David Marwell, personal interview.
216 The youngsters affectionately called him "*titio*" . . . Gypsies of Auschwitz: OSI archives.

10 The Scholar and the Preacher

219 The 1972 Munich Olympics brought back . . . host to the entire world: Mengele correspondence, undated letter to Hans Sedlmeier, courtesy of *Stern*.
219 In a letter to Sedlmeier . . . coverage in Brazil: Ibid.
219 "I am an old athlete" . . . his boyhood friend: Ibid.
219 "Of course, this is not the proper method either," he lamented in his diary: Mengele diary, September 5, 1972.
219 Any new operations would be sure to put them at risk: See *Ma 'Ariv* November 5, 1982, p. 2.
219 In Paraguay, for instance . . . didn't appreciate questions about Mengele: Ibid. Also, Enrique Matalon, prominent Jewish community leader of Asunción, personal interview, Asunción, Paraguay, November 1984.
219 He would hold frequent . . . to finding Mengele: See "Paraguay Keeps Mengele Comfortable—Wiesenthal," *Jerusalem Post*, August 18, 1977.
220 A newsletter he published . . . on Mengele's supposed whereabouts: See "Bulletin of Information" published by Wiesenthal's Documentation Center, Vienna, No. 12, January 1972, p. 3. Under heading number 5, "Dr. Josef Mengele," Wiesenthal wrote as follows: "This criminal, wanted throughout the world, stayed in Torremolinos, Spain, in March 1971, according to our reliable information. Our informant was able to take down the license number of the car which Mengele drove; it was a German number."
220 It was only a matter of time . . . Wiesenthal would confidently proclaim: See Bulletin 21, January 31, 1981. "Now that we know somewhat more about Mengele's friends and his circle of acquaintances, we hope for success at last in 1981," Wiesenthal wrote.
220 Wiesenthal promoted . . . the victim of a mysterious "accident": Simon Wiesenthal, *The Murderers Among Us*, p. 159.
220 As a result of his obsessive worrying . . . hairballs in his stomach: David Marwell, OSI, personal interview.
220 Terrified of being discovered . . . using his alias, Peter Hochbichler: Ibid.
220 When the doctor ordered extensive X rays . . . to all of them: Ibid.
220 To the doctor's surprise . . . extra copies of the film: Ibid. Obviously, the search

for modern X rays of Mengele became extremely important for the Justice Department in 1985, as it sought to verify whether the skeletal remains found in Brazil were indeed those of the Auschwitz death doctor. But the agency came up empty-handed.

220 The surgery was successful . . . of his medical records: Ibid.

221 And Bossert sent him a note . . . "understandable to idiots": Mengele correspondence, undated letter from Bossert to Mengele, courtesy of *Stern*.

221 A special top-secret arrangement . . . to and from Mengele: Posner and Ware, *The Complete Story*, pp. 232–33.

221 Mengele would send his letters . . . to whom they were addressed: Ibid.

221 "I always feel hurt" . . . complained to someone code-named "Kitt": Mengele correspondence, undated letter, courtesy of *Stern*.

221 "Your neglect is a source of great pain to me": Ibid.

221 "I heard a rumor" . . . he writes to Sedlmeier: Mengele correspondence, undated letter, courtesy of *Stern*.

221 "If I had to choose again, I would choose the same profession," the Auschwitz doctor blithely observes: Ibid.

223 Sixty-four-year-old Josef Mengele . . . fifty-one-year-old Wolfgang Gerhard: *Newsweek*, June 24, 1985, p. 44.

223 "You are not in any position" . . . status as a wanted man: Mengele diary, January 18, 1975, courtesy of *Stern*.

223 She was going to sell the farm . . . for the death-camp doctor: *New York Times*, June 9, 1985, p. 16.

223 That same year, Mengele had moved into his own house: Ibid.

223 As a final favor . . . rented it to Mengele: Posner and Ware, *The Complete Story*, p. 243.

223 A small stroke . . . his sense of helplessness: David Marwell, OSI, personal interview.

223 One Christmas, when Rolf . . . and his stepson, Karl Heinz: Inge Byhan, *Bunte* V, handwritten translation by Elizabeth Bergmann, p. 2.

223 "The letter from Karl Heinz . . . from Rolf was too factual": Ibid.

224 Sharp alternances in sentiment . . . Mengele's correspondence with his son: Ibid.

224 "Rational observance" . . . to select a mate: Mengele correspondence, undated letter to Rolf Mengele, courtesy of *Stern*.

224 His letters suggest he thought his son was shallow and superficial: Inge Byhan, *Bunte* V, p. 1.

224 "You and your generation" . . . wrote at one point: Mengele correspondence, as reviewed by Professor Zdenek Zofka.

224 Mengele faulted Rolf for . . . money, cars, and other comforts: Ibid.

225 "There was an air raid" . . . Mengele recalled: Mengele letter to Rolf, undated, courtesy of *Stern*.

225 "The worrying and responsibility" . . . he wrote: Ibid.

226 "I accept that you lack interest" . . . no claim to the family fortune: Mengele letter to Rolf, undated, courtesy of *Stern*.

226 Mengele went on to point out how foolish . . . "three years of training": Ibid.

226 Deeply upset, he remarked . . . "to express my feelings": Ibid.

226 "I was very hurt": Ibid.
226 "The doctorate was the only desire I ever had for you": Ibid.
227 The relationship between father and son . . . good Aryan stock: Professors Zdenek Zofka and Uwe-Dietrich Adam, personal interviews.
227 Dr. Mengele was delighted . . . choice of mate: Ibid.
227 Mengele was both fascinated . . . I had married a twin: Inge Byhan, *Bunte* V, p. 14.
227 "For the first time" . . . he later exulted: Ibid.
227 "This movement north . . . from this new match": Ibid.
227 "Even the characteristic of Almuth" . . . Mengele noted: Ibid.
229 So as not to endanger . . . a false passport: David Marwell, OSI, personal interview.
229 In May 1977 . . . to São Paulo: Inge Byhan, *Bunte* V, p. 2.
229 Although he was employing an alias . . . authorities to his father: David Marwell, OSI, personal interview.
230 According to Rolf, the reunion was tender and sentimental: Inge Byhan, *Bunte* V, pp. 13–14.
230 In interviews given years later . . . warm and loving: Ibid.
230 "You surely do not believe" . . . about Auschwitz: Ibid.
230 It was then that Dr. Mengele swore . . . "personally" killed anyone: Ibid.
231 In an effort to woo Rolf . . . slept on the floor: Posner and Ware, *The Complete Story*, p. 277.
231 Mengele had even used . . . his son's future bride: Inge Byhan, *Bunte* V, p. 14.
231 The daily entries . . . crammed with increasingly mundane details: Mengele 1978 diary, courtesy of *Stern*.
232 Restless and unable to sleep, he read late into the night: Ibid.
232 The object of his affections . . . forty years his junior: *Newsweek*, June 24, 1985, p. 44.
232 Elsa has recalled that . . . with chaste ardor: Ibid.
232 As their relationship deepened . . . with him: Ibid.
232 In his notebooks . . . "proper" conduct between men and women: Mengele notebooks, undated, courtesy of *Stern*.
232 Like a preacher . . . "perversion": Ibid.
232 "Sexual barriers prevent chaos" . . . in one entry: Ibid.
232 But he also observed . . . "expression of passion": Ibid.
232 But when she realized it would never be . . . to someone else: *Newsweek*, June 24, 1985, p. 45.
232 Mengele was devastated: Ibid.
233 As Wolfram later confessed . . . even though they liked him: Letter from Wolfram Bossert to Karl Heinz Mengele, April 1979, courtesy of *Stern*.
233 He "demanded conformity of everyone," Bossert wrote: Ibid.
233 "Whoever didn't manage . . . own sense of identity": Ibid.
233 That was why . . . impossible: Ibid.
233 "It was important for us to be somewhat distant": Ibid.
233 Birthdays continued to go unnoticed . . . "much more bearable": Mengele diary, March 16, 1975.

233 Another year, Bossert surprised him . . . "unfortunately, no letters": Mengele diary, March 16, 1979.
233 Once, Mengele asked Gerhard . . . what a fine person he was: Mengele correspondence, letter to Wolfgang Gerhard, February 24, 1972.
233 "I would like for them" . . . before the planned encounter: Ibid.
233 "This way, we can avoid . . . and the negative attitudes": Ibid.
234 Mengele's last complete diary . . . any of the others: Mengele diary, 1978, courtesy of *Stern*.
234 He condemns the local obsession with South American soccer scores: Mengele diary, June 8, 1978.
234 He rails against a jazz festival . . . "musical schizophrenia": Mengele diary, September 17, 1978.
234 With contempt . . . the Galapagos Islands: August 9, 1978.
234 With finality, he labels . . . "politics": Mengele diary, January 6, 1978.
234 He watches the heavyweight title fight . . . "stupefied mass culture": Mengele diary, September 15, 1978.
234 Only one popular icon . . . starring Marilyn Monroe: Mengele diary, May 19, 1978.
234 "I am losing hope" . . . in one entry: Mengele diary, September 9, 1978.
234 It is the fifth week . . . without sleep: Ibid.
235 Although he was ailing . . . had begun so many years ago: Mengele diary, January 10, 11, 1978.
235 Looking back with longing . . . had been destroyed: Mengele notebooks, courtesy of *Stern*.
237 One day, as he listened . . . Opus 6: Mengele diary, May 23, 1978.
237 As a little boy . . . why she was so sad: Ibid.
237 "Boy, you cannot understand that yet," his mother had replied: Ibid.
237 "I understand better now, Mother," he wrote in his diary: Ibid.
237 Finally, at three o'clock in the morning . . . beach house: Wolfram Bossert, "The Last Day," Letter from Wolfram Bossert to members of the Mengele family, undated.
238 As the Bosserts would later recall . . . terrible lot in life: Ibid.
238 On and on he rambled . . . nonexistent cultural life: Ibid.
238 "This is what happens" . . . to the children: Ibid.
238 "I am not really that old," sixty-eight-year-old Mengele scoffed indignantly: Ibid.
238 According to the Bosserts . . . suffered a stroke: Ibid.
238 "Come back, Uncle," he cried, "the ocean is pulling you in": Ibid.
238 Their father tried in vain to revive him, but it was too late: Ibid.
238 Dr. Mengele was dead: Ibid.
238 A lifeguard on the scene . . . in front of him: Ibid.
238 Liselotte saved the day . . . "made him age remarkably": Ibid.
238 At the beach . . . the Angel of Death: Ibid.
239 A wreath of red roses . . . mercy to the departed: Ibid.

11 The Burial of the Dead

240 In December 1979, Rolf Mengele says . . . he had hardly known: Inge Byhan, *Bunte* III, p. 1.
240 It had taken him months to save up money for the air fare: Ibid.
241 Rolf recalled thinking . . . small, forlorn gravesite: Ibid.
241 What Rolf cared about most . . . barely legible script: Ibid.
241 Mengele's son also combed . . . enigma that was his father: Ibid.
241 Rolf and other family members . . . had drowned on a Brazilian beach: *Washington Post*, June 13, 1985, p. 5.
241 Rolf Mengele himself insisted that his family . . . protected his father over the years: Ibid. Rolf said he remained silent "out of consideration for the people who were in contact with my father for the last thirty years."
241 Intensely private, Martha declines . . . her life with Mengele: Martha Mengele letter to Lucette Matalon Lagnado, September 20, 1985.
242 But through the years . . . as a young girl: Hermann Abmayr interview with Irene Schoenbein, date unknown. See also letter from Irene Schoenbein to S. Jones and K. Rattan, April 14, 1984.
242 Rolf claims he felt strangely relieved after his father's death: Inge Byhan, *Bunte* III, p. 2.
243 Convinced that the government . . . somewhere in the vicinity of Asunción: Ambassador Robert White, "Where is Mengele?" Cable from U.S. embassy in Asunción to State Department, April 10, 1979.
243 He took to wandering alone . . . the progress of his "hunt": Ibid. Also, Robert White, personal interview, 1985.
243 In a surprise move . . . after granting it to him: Cable, U.S. embassy in Asunción to State Department, August 9, 1979.
243 It was a gesture . . . particularly those in Washington: Ibid.
243 The embassy in Asunción had learned . . . on a Delta Airlines flight: Secret cable, U.S. embassy in Asunción to State Department, August 28, 1979.
243 Although information was sketchy . . . under his name: Ibid.
243 The bizarre report . . . to arrest Mengele when he landed: Secret cable, State Department to U.S. embassy Asunción, August 28, 1979.
243 It was Vance who gave . . . when he arrived in Miami: Secret Cable, State Department to U.S. embassy in Asunción, August 28, 1979, signed "Vance."
243 He continued to tantalize the world with fantastic tales of the Auschwitz doctor: See Wiesenthal bulletins 21, January 31, 1981; 22, January 31, 1982; 23, January 31, 1983.
244 In 1983, the first national conference . . . of Hitler's death camps: *Virginian Pilot*, April 12, 1983, p. 1.
244 Struck by the poignancy of Eva's letter . . . the Indiana housewife: Lucette Matalon Lagnado interviews with Eva Mozes, Winter 1984.
245 With the help of a relative . . . experiments to contact her: *Ma'Ariv*, February 6, 1984. Also, Miriam Mozes, personal interview, March 1985.
246 *Parade* magazine's publication . . . spotlight on CANDLES: Jack Anderson, "The Twins of Auschwitz Today," *Parade* magazine September 4, 1984.
248 After Auschwitz, the twins . . . Jewish Holocaust victims: Thomas L. Friedman,

"Jerusalem Listens to the Victims of Mengele," *New York Times*, February 7, 1985, p. 1.

249 General Telford Taylor . . . for the trial: Ibid.

249 And so was Gideon Hausner . . . Adolf Eichmann: Ibid.

249 Simon Wiesenthal . . . honor on the podium: Eva Mozes, personal interview.

249 He had chuckled: "Now, I have work for twenty years": Friedman, *New York Times*, February 7, 1985, p. 12.

249 At the close . . . "crimes against humanity": Statement of tribunal, courtesy of CANDLES.

250 The West German and Israeli governments . . . to find Mengele: Neal Sher, director, OSI, personal interview.

250 As neither wanted to share . . . intermediary: Ibid.

250 Yet when officials . . . amounted to very little: Ibid.

250 There was no recent photograph known for certain to be of him: Ibid.

250 Although he was said . . . he could be anywhere: David Marwell, OSI, personal interview.

250 Mengele was spotted in Las Vegas . . . a marimba band in California: Ib d.

250 At one point . . . had checked in: Ibid.

251 They hurried up . . . Lolo's flustered son: Ibid.

251 But files at the National Archives . . . when it came to Mengele: Dr. Robert Wolfe, director, Captured German Records, U.S. National Archives, personal interview.

251 But there was nothing . . . and the American postwar government: David Marwell, OSI, personal interview.

251 A surprise search by German authorities . . . ostensibly from Mengele: *Washington Post*, June 13, 1985, p. 35.

251 Searchers also found a notebook . . . Bosserts and the Stammers: *Washington Post*, June 11, 1985, p. 3.

252 Pressed to come up . . . and incomplete data: Letter from Eli Rosenbaum, Counsel, World Jewish Congress, to unnamed recipient, August 29, 1985, p. 2.

252 Within three days . . . Dr. Josef Mengele: Ibid. p. 6.

252 Draping herself in the soft white shawl . . . with the Auschwitz doctor: *Time*, June 24, 1985, pp. 40–41.

252 There are suspicions that Mengele's son sought to profit from the publicity surrounding his father's death: Jack Anderson and Joseph Spear, "Son Trades on Father's Infamy," *Syndicated Washington Merry-Go-Round Column*, August 29, 1985.

252 He is even believed to have sought a Hollywood contract to make a movie about his father: The reports are vague, but officials at the Simon Wiesenthal Center assert that Rolf tried to get a movie deal, which they quashed the instant they learned of it.

252 Under oath, they swore . . . and medical records: Eli Rosenbaum, deputy director, Justice Department Office of Special Investigations, personal interview. (Rosenbaum, who had first raised questions about the pathologists' work while at the World Jewish Congress, continued to press the matter when he joined OSI.)

252 But an attorney at the World Jewish Congress raised questions about the doctors' findings: Ibid.

252 The virulent bone disease . . . ought to have revealed to Congress: Ibid.

252 In Washington, then—attorney general Edwin Meese announced that the U.S. search for Dr. Mengele was over: "U.S. Agrees Mengele Is Dead," Associated Press, August 3, 1986.

253 Behind the scenes . . . close their Mengele files: David Marwell, OSI, personal interview.

253 Nor was Germany able . . . the remains were those of the Auschwitz doctor: Ibid.

253 In Washington, Frankfurt, and Jersualem . . . osteomylitis: Ibid.

253 Anxious to settle the question . . . traces of the osteomylitis: Ronald Soble, "U.S. Secrecy Keeps Cloud of Doubt Over Mengele Death," *Los Angeles Times*, March 3, 1989, p. 26.

253 But he, too . . . of the illness: Ibid.

253 With the exception of the dental forensic scientist . . . of the skeleton: Neal Sher, OSI, personal interview, September 1990.

253 Only with the discovery . . . to the Justice Department: Dr. Lowell Levine, personal interview, September 1990.

253 One problem hampering U.S. officials . . . at the same time: Neal Sher, OSI, personal interview.

254 After the discovery of the bones . . . of his own: Menachem Russek, personal interview, Tel Aviv, Israel, May 1988.

254 A colonel in the Israeli police . . . close their Mengele files: Ibid.

254 Russek also spent months . . . belonged to Mengele: Ibid.

254 Russek concluded the body . . . world off the scent: Ibid.

254 He prepared a scathing . . . "the Angel of Death, Dr. Josef Mengele": Menachem Russek, Mengele report, as shown to Lucette Matalon Lagnado, May 1988, Tel Aviv, Israel.

255 In 1987, the German government . . . to erase the past: Eva Mozes, CANDLES president, personal interview, September 1990.

Selected Bibliography

I. Books and Dissertations

ALLEN, WILLIAM SHERIDAN. *The Nazi Seizure of Power: The Experience of a Single German Town 1930–1935*. Chicago: Quadrangle Books, 1965.

ARENDT, HANNAH. *Eichmann in Jerusalem: A Report on the Banality of Evil*. New York: Penguin, 1963.

BERNARDAC, CHRISTIAN. *L'Holocauste Oublié*. Paris: Editions France-Empire, 1979.

DABRINGHAUS, ERHARD. *Klaus Barbie: The Shocking Story of How the U.S. Used This Nazi War Criminal as an Intelligence Agent*. Washington, D.C.: Acropolis Books, 1985.

DAWIDOWICZ, LUCY S. *The War Against the Jews 1933–1945*. New York: Bantam Books, 1976.

FARAGO, LADISLAS. *Aftermath: Martin Bormann and the Fourth Reich*. New York: Simon and Schuster, 1974.

FÉNELON, FANIA. *The Musicians of Auschwitz*. London: Michael Joseph, 1986.

HART, KITTY. *Return to Auschwitz*. New York: Atheneum, 1982.

HAREL, ISSER. *The House on Garibaldi Street*. New York: Viking Press, 1975.

HAUSNER, GIDEON. *Justice in Jerusalem*. New York: Harper and Row, 1966.

HITLER, ADOLF. *Mein Kampf*, trans. Ralph Manheim. Boston: Houghton Mifflin, 1943.

HOSS, RUDOLF, Pery Broad, and Johann Paul Kremer. *KL Auschwitz Seen by the SS.* New York: Howard Fertig, 1984.

KEVLES, DANIEL J. *In the Name of Eugenics: Genetics and the Uses of Human Heredity.* New York: Alfred A. Knopf, 1985.

LENGYEL, OLGA. *Five Chimneys.* New York: Howard Fertig, 1983.

LIFTON, ROBERT JAY. *The Nazi Doctors: Medical Killing and the Psychology of Genocide.* London: Macmillan, 1986.

MOSSE, GEORGE L. *The Crisis of German Ideology: Intellectual Origins of the Third Reich.* New York: Grosset and Dunlap, 1964.

NEWTON, RONALD C. *German Buenos Aires 1900–1933: Social Change and Cultural Crisis.* Austin: University of Texas Press, 1977.

MÜLLER-HILL, BENNO. *Todliche Wissenschaft: Die Aussonderung von Juden, Zigeunern und Geisteskranken 1933–1945.* Hamburg: Rowohlt, 1984.

NOMBERG-PRZYTYK, SARA. *Auschwitz: True Tales from a Grotesque Land,* ed. Eli Pfefferkorn and David H. Hirsch. Chapel Hill: University of North Carolina Press, 1985.

NYISZLI, DR. MIKLOS. *Auschwitz: A Doctor's Eyewitness Account.* New York: Frederick Fell, 1960.

PAGE, JOSEPH. *Perón: A Biography.* New York: Random House, 1983.

PERL, GISELLA. *I Was a Doctor in Auschwitz.* New York: International Universities Press, 1948.

POSNER, GERALD L., and JOHN WARE. *Mengele: The Complete Story.* New York: McGraw-Hill, 1986.

PRIDHAM, GEOFFREY. *Hitler's Rise to Power: The Nazi Movement in Bavaria 1923–1933.* (New York: Harper and Rowe, 1973).

SPEER, ALBERT. *Inside The Third Reich—Memoirs,* trans. Richard and Carla Winston. New York: Collier Books, 1970.

STEVEN, STEWART. *Spymasters of Israel.* New York: Macmillan, 1980.

THOMANN, KLAUS-DIETER. *Medizin im Faschismus.* Berlin: Veb Verlag Volk und Gesundheit, 1985.

WEINREICH, MAX. *Hitler's Professors.* New York: Yiddish Scientific Institute—YIVO, 1946.

WIESEL, ELI. *Night.* New York: Bantam Books, 1982.

WIESENTHAL, SIMON. *The Murderers Among Us.* New York: McGraw-Hill, 1967.

ZOFKA, ZDENEK. *Die Ausbreitung des Nationalsozialismus auf dem Lande: Eine Regionale Fallstudie zur Politiscehn Einstellung der Landbevolkerung in der Zeit des aufstiegs und der Machtergreifung der NSDAP 1928–1936.* (Published Dissertation) Munich: Stadtarchiv Munchen, 1979.

II. *Newspapers, Journals, and Periodicals*

ANDERSON, JACK. "The Twins of Auschwitz Today." September 4, 1984.

ANDERSON, JACK, and JOSEPH, SPEAR. "Son Trades on Father's Infamy." *Washington Merry-Go-Round* syndicated column, August 29, 1985.

BLUMENTHAL, RALPH. "Mengele Family Members Are United In Their Silence." *New York Times,* June 11, 1985, p. 4.

BOOZER, JAC S. "Children of Hippocrates: Doctors in Nazi Germany." *Annals of the American Academy of Political and Social Science*, Vol. 450, July 1980, p. 84.

BYHAN, INGE. Mengele Series, Parts I–V. *Bunte*, summer 1985. Archives of Burda Publications, Munich, Germany.

DROZDIAK, WILLIAM. "Mengele Family Plans to Issue Statement." *Washington Post*, June 11, 1985.

FRIEDMAN, THOMAS L. "Jerusalem Listens to the Victims of Mengele." *New York Times*, February 7, 1985.

FRIEDRICH, OTTO. "The Kingdom of Auschwitz." *Atlantic Monthly*, September 1981, pp. 30–60.

HELMUTH, G. "*Die Vererbung Als Biologischer Vorgant*." *Der Weg*, undated (believed to be from 1953), courtesy of *Stern*.

HOUSE, RICHARD. "Woman Says Mengele Spent 1961–1979 in Brazil." *Washington Post*, June 9, 1985, pp. 1, 28.

HOUSE, RICHARD. "Probers Say Proof Mounts That Mengele Was in Brazil." *Washington Post*, June 11, 1985, pp. 1, 14.

Jerusalem Post, July 10, 1964 (Frankfurt trial brief).

Jerusalem Post, August 18, 1964 (Frankfurt trial brief).

LEWIS, PAUL. "Couple Will Face a German Inquiry on Aid to Mengele." *New York Times*, June 10, 1985.

MARKHAM, JAMES M. "Mengele 'Double' Called Fervid Nazi." *New York Times*, June 13, 1985.

MATHEWS, JAY. "Researchers Find Signs U.S. Knew of Mengele's Postwar Hideout." *Washington Post*, March 15, 1985, p. 24.

MEYER, ERNIE. "Mother Testifies How She Killed Her Baby." *Jerusalem Post*, August 2, 1985.

"Paraguay Keeps Mengele Comfortable—Wiesenthal," *Jerusalem Post*, August 18, 1977.

Reuters News Service. "Bormann Said Seen With Mengele." *Jerusalem Post*, July 19, 1964.

Reuter News Service. "Bonn Told: Mengele is citizen of Paraguay." *Jerusalem Post*, August 21, 1964.

RIDING, ALAN. "Woman Says Man Who Admitted He Was Dr. Mengele Died in 1979." *New York Times*, June 9, 1985, p. 16.

SHUB, ANATOLE. (Missing Title.) *Washington Post*, August 20, 1965, p. 13.

SOBLE, RONALD. "U.S. Secrecy Keeps Cloud of Doubt Over Mengele Death." *Los Angeles Times*, March 3, 1989, p. 26.

THOMANN, KLAUS-DIETER. "Racial Hygiene and Anthropology: Professor Verschuer's Twin Careers." *Frankfurter Rundschau*, May 21, 1985, p. 8.

WATSON, RUSSELL, MAC MARGOLIS, and ANDREW NAGORSKI. "Who Helped Mengele?" *Newsweek*, June 24, 1985, pp. 40–45.

WHITEFIELD, MIMI. "Reports Say Alleged Nazi Lived Life of Ease." *Boston Globe*, June 9, 1985, pp. 1, 20.

"The Mengele Mystery." *Time*, June 24, 1985, p. 40.

"U.S. Agrees Mengele Is Dead." Associated Press, August 3, 1985.

III. *Personal Papers of Dr. Josef Mengele*

Diaries of Dr. Josef Mengele, 1962–1978, courtesy of *Stern*.
Letters of Dr. Josef Mengele to family and friends, dates unknown, courtesy of *Stern*.
Notebooks of Dr. Josef Mengele, dates unknown, courtesy of *Stern*.

IV. *Archival Material*

Unpublished documents and reports in possession of:
Auschwitz Museum, Auschwitz, Poland.
Berlin Document Center, Berlin, Germany.
B'nai B'rith Anti-Defamation League archives, New York.
Polish Commission for the Investigation of War Crimes, courtesy Polish archives, Warsaw, Poland.
U.S. Justice Department, Office of Special Investigations files, Washington, D.C.
U.S. National Archives, Washington, D.C.
U.S. State Department, Washington, D.C.

INDEX